EMISSION CONTROL FROM INDUSTRIAL BOILERS

EMISSION CONTROL FROM INDUSTRIAL BOILERS

Edited by

HOWARD E. HESKETH, Ph.D., P.E., D.E.E.
Consultant/Professor
Southern Illinois University, Carbondale, IL

FRANK L. CROSS, JR., P.E., D.E.E.
Consultant
Harding Lawson Associates, Orlando, FL

JOHN T. QUIGLEY, Ph.D., P.E., D.E.E.
Professor
University of Wisconsin, Madison, WI

CRC Press
Taylor & Francis Group
Boca Raton London New York

CRC Press is an imprint of the
Taylor & Francis Group, an **informa** business

Emission Control from Industrial Boilers

First published 1995 by Technomic Publishing Company, Inc.

Published 2018 by CRC Press
Taylor & Francis Group
6000 Broken Sound Parkway NW, Suite 300
Boca Raton, FL 33487-2742

ISBN 13: 978-1-56676-182-6 (hbk)

**Visit the Taylor & Francis Web site at
http://www.taylorandfrancis.com**

**and the CRC Press Web site at
http://www.crcpress.com**

Main entry under title:
  Emission Control from Industrial Boilers

Bibliography: p.
Includes index p. 189

Library of Congress Catalog Card No. 94-61079

Table of Contents

Preface

THE Clean Air Act Amendments (CAAA) of 1990 significantly affect commercial and industrial combustion devices such as boilers, incinerators and other burners. Under the new emission regulations already promulgated and those being developed, compliance will require improved equipment, more detailed operator training, new permits, more complex monitoring and reporting, as well as other requirements. All emissions must be considered, e.g., particulates and gases (acid, organic, hazardous, NO_x, ozone).

Many industrial boiler plants have been retrofitted to change fuel and/or combustion operating conditions as a means to meet new air pollution control requirements. New regulations will continue to be developed by the CAAA of 1990 that will require changes to other boilers and combustion systems. This book is intended to acquaint industry with the equipment and operating options that are available to reduce emissions while controlling costs.

Specific topics are addressed, including regulatory requirements, boiler and burner equipment retrofits, combustion modification, air emission control and monitoring equipment selection, maintenance and cost. The twelve chapters of this book are written by seven different authors. The authors use fifty-two figures and forty-four tables to help explain the written text and to make it more interesting and useful to you, the reader.

HOWARD E. HESKETH, PE
FRANK L. CROSS, JR., PE
JOHN T. QUIGLEY, PE

ix

Acknowledgements

THE initial assemblage of this material occurred as the result of a Continuing Engineering education course at the University of Wisconsin at Madison, under the direction of Dr. J. T. Quigley. We appreciate the help of Marisue K. Quigley, who arranged with the authors to prepare their respective chapters.

Special thanks go to the Southern Illinois University at Carbondale, Department of Mechanical Engineering and Energy Processes, for help in preparing the manuscript. This includes Amy Dunn (for devotedly performing the word-processing chores), Judy Cockrum (for supervising the office activities) and to the office assistants (for making copies and providing other help).

Finally, a great expression of gratitude goes to Scott Lane, Mechanical Engineering graduate student at Southern Illinois University, for his many hours of dedicated proofreading.

1

The Clean Air Act

FRANK W. SHERMAN
HOWARD E. HESKETH

1.1 BACKGROUND

A IR pollution regulations may be set by federal, state or local regulatory bodies. Regulations set by any of these bodies must be at least as strict as those established by the next higher level of government. Federal regulations set by Congress are published daily in the *Federal Register.* Those are then, in effect, laws.

These regulations are codified annually into the Code of Federal Regulations, CFR. Title 40 of the CFR is "Protection of Environment." This contains all the EPA regulations and covers rule making and policies of the Administrator. Title 40 part numbers and subchapter topics are

- 40 CFR 1–51 (Air Regs, Grants/Federal Assis.)
- 40 CFR 52–99 (Air Programs)
- 40 CFR 100–149 (Water Programs)
- 40 CFR 150–189 (FIFRA-Pesticide Programs)
- 40 CFR 190–259 (RCRA, Radiation, Ocean Dumping, Solid Wastes)
- 40 CFR 260–299 (RCRA, Hazardous Wastes)
- 40 CFR 300–399 (CERCLA/Superfund/SARA Title III)
- 40 CFR 400–699 (Water Effluent Guidelines and Standards)
- 40 CFR 700–END (TSCA)

The Clean Air Act (CAA) of 1963 initiated most of the air pollution control activities in the U.S. This act is amended periodically by Congress. For example, it was significantly amended in 1965, 1966, 1967 (Federal Air Quality Act), 1968, 1970, 1977, 1987 (Amendments to Criteria Pollutant Values) and 1990.

1

The Clean Air Act provisions for cleaning the air are based upon the establishment of Air Quality Standards for several pollutants. Since 1970, the EPA has established National Ambient Air Quality Standards (NAAQS) for five "criteria" pollutants: ozone, carbon monoxide, particulate matter, nitrogen oxides and sulfur dioxide. Lead and respirable particulate (i.e., PM_{10}—particulate matter with a mean diameter less than ten microns) were added later. The NAAQS for each of these pollutants was based upon a "criteria document" which summarized the latest research concerning air quality levels at which human health (primary standard) and welfare (secondary standard) are seriously impacted. In adopting the health, or primary standards, EPA is not allowed to consider economic or technical feasibility of emission control requirements. Air quality standards are to be reviewed every five years by EPA, but they have not, in fact, been reviewed with that regularity.

The 1990 Clean Air Act Amendments (CAAA) were the first major revision to that Act in thirteen years, and upon their adoption introduced significant changes in the way the Act allows—indeed requires—EPA to regulate industry and work with the states, and the way it requires industry to cope with regulatory issues. It uses the carrot and stick approach to environmental reform with bigger carrots and sticks than had been used under the Amendments of 1977. The new Act is detailed and complicated. The regulations which are and will be promulgated pursuant to its many provisions will be no less complicated. A broad understanding of the implications of this legislation is important for any professional who has environmental responsibilities at the plant level or above.

The 1990 amendments are the last, to date, for the CAA and were signed into law by the president in November 1990. This is one of the most significant pieces of environmental legislation ever enacted, and it will generate hundreds of regulations during the 1990 decade and on into the twenty-first century.

1.2 COEXISTING WITH THE CAAA

Industries with air emissions need to determine the applicability of these new regulations to their processes and operations. For example, it is necessary to identify hazardous air pollutants (HAPs) requirements and to evaluate if volatile organic compounds (VOCs) requirements are overlapping or additional. Industry must be cognizant of the new operating permits and the cost of control options.

In order to comply with the regulations, industries could use a three-step process:

(1) Develop emission inventory data for all air emissions. This includes

identifying the specific units, identifying the emissions by chemical compounds, and accurately establishing quantities. Be certain to document how these data are obtained.

(2) Determine applicable requirements. Use information, such as that provided in the rest of this chapter, to show (based on your emission inventory) if a certain regulation, or regulations, are triggered. If so, study that regulation in greater detail. Emission source geographical location is also critical to include as different requirements exist for attainment and non-attainment areas. Be sure to note that in the 1990 CAAA are items such as accidental releases, workers' safety under Occupational Safety and Health Administration (OSHA), source reports and monitoring records.

(3) Develop a plan of action to achieve compliance. Using the information compiled under (1) and (2) above, prepare a logical sequence for action to secure operating permits, obtain proper control equipment and/or process procedures, develop worker training programs, manage data effectively, strive for good community public relations and initiate quality reporting procedures.

The following material of Chapter 1 is a description of *some* of the requirements in the 1990 CAAA, which may apply to an operator of a large stationary source such as an industrial boiler or the plant at which it is located. To present a complete synopsis of the Clean Air Act Amendments as they apply to stationary sources is beyond the scope of this book, and the following is not intended to be legal advice. It is intended, however, to provide the reader with an understanding of the breadth and complexity of the federal regulatory process, which applies to air emission sources, so that a plant engineer can appreciate the need for a thorough understanding of its requirements and implications.

1.2.1 FEDERAL CLEARINGHOUSE

Most major source construction projects will require that the source in question demonstrates a degree of emission control or efficiency that has been demonstrated by existing technology, and that source may be either subject to a BACT, LAER, MACT or ACT standard. EPA has established a clearinghouse to collect such information on new source technology and what constitutes its latest findings in each control technology area. Called the Technical Transfer Network (TTN), this computer-accessible clearinghouse is required by the Clean Air Act to provide information on RACT, BACT and LAER, and related air quality planning topics. Further information on this data source can be obtained from the Environmental Protection Agency Office of Air Quality Planning and Standards (OAQPS), Research Triangle Park, North Carolina.

1.3 THE CAA AMENDMENTS OF 1990

On November 15, 1990, President Bush signed into law (Public Law No. 101-549) amendments to the Clean Air Act, 42 U.S.C. 7401. The 1990 Amendments are extremely lengthy and are both far-reaching and complex. In addition, many new rules and procedures will need to be developed. This chapter summarizes some of the most significant aspects of the new CAAA. The 1990 CAAA are broken down into sections called Titles. The titles of interest relative to boiler systems and presented in this chapter are

- Title I: Non-Attainment Provisions
- Title III: HAPs
- Title IV: Acid Rain
- Title V: Operating Permits
- Title VI: Enforcement

1.4 TITLE I: NON-ATTAINMENT PROVISIONS

1.4.1 INTRODUCTION

The Amendments retain the current non-attainment program, which requires the United States Environmental Protection Agency (EPA) to establish National Ambient Air Quality Standards (NAAQS) for criteria pollutants [e.g., sulfur dioxide (SO_2), particulate matter (PM_{10}), nitrogen dioxide (NO_2), carbon monoxide (CO), ozone and lead]. NAAQS represent the maximum allowable concentration of the pollutants allowed in the ambient air. States bear the responsibility of ensuring that NAAQS are met. State Implementation Plans (SIPs) containing emission standards for specific sources are developed by states to achieve and maintain NAAQS. Prior to the Amendments, SIPs were required only to regulate major new sources and major modifications. When an area fails to meet a NAAQS, it is considered a "non-attainment area." More stringent control requirements, designed to achieve attainment, must be applied in non-attainment areas.

The USEPA exercises three primary programs for the control of stationary source emissions:

- New Source Performance Standards

- Prevention of Significant Deterioration (not discussed in this chapter)
- Non-attainment Areas Requirements

1.4.2 NEW SOURCE PERFORMANCE STANDARDS (NSPS)

NSPS have been used for several years as a means of ensuring that modern facilities or major modifications of existing sources utilize a standard of emission control that is uniformly applied in all areas, irrespective of air quality considerations. The standards are based upon technical and economic considerations that apply to new plant construction; such standards have served two purposes: (a) they define a benchmark level of control that is, in most cases, environmentally sound, and (b) they remove any incentive for a facility to build new plants in areas that have less stringent air pollution control requirements. New Source Performance Standards are codified in federal regulations. However, a new source seeking to build a major source in the United States will generally be subject to a new source review process, which will vary depending upon whether the source will be located in an attainment or non-attainment area. In attainment areas, Prevention of Significant Deterioration (PSD) rules apply. In non-attainment areas, a new source review process is required, which ensures that the proposed facility can be built within the limitations of an approvable State Implementation Plan (SIP). Each of these cases is further discussed below.

1.4.3 ATTAINMENT AREAS

Construction of a major new facility in an attainment area is governed by the rules on Prevention of Significant Deterioration (PSD). These rules establish prescribed increments of air quality which a new source is allowed to use, provided it meets certain other requirements, chief of which is the use of Best Available Control Technology (BACT). BACT is a level of control that may be of higher stringency (never less) than a NSPS, and is based upon technology which is currently available in a commercial form and represents the state-of-the-art of that technology. Since the state-of-the-art is continuously being advanced, BACT is a changing standard.

1.4.4 NON-ATTAINMENT AREAS

A non-attainment area may be defined in many ways, from a street corner for carbon monoxide to a county or larger for ozone. Although the standards upon which non-attainment areas are determined are derived from federal standards, it is the responsibility of the respective states to develop the strategies that will result in attainment of the air quality standards in these areas. These state proposals for achieving air quality standards are called State Implementation Plans, or "SIPs." A SIP must be approved by USEPA before it can be implemented. Several of the important components of any approvable SIP are

- adequate air quality monitoring data
- legal authority to enforce standards
- adequate state resources
- an approvable permit fee schedule
- technical support for control measures
- an approvable air quality analysis
- evidence of public participation in the planning process

In cases where the state does not act within twenty-four months to submit an approvable plan, EPA may adopt its own plan, called a Federal Implementation Plan, or FIP. As of this writing only two states have non-attainment areas that are operated under an FIP—California and Illinois.

Prior to making any long-term investment in a major source construction project, an environmental engineer should contact affected state or local air quality planning officials to determine the status of the SIP or FIP in his state.

1.4.5 OZONE NON-ATTAINMENT

Unlike most other pollutants, ozone non-attainment may fall into one of several categories—each category represents a different level of air quality deterioration. The higher the ozone pollution level, the longer the time that is allowed for the non-attainment area to come into compliance. Table 1.1 presents these classifications of non-attainment and some of the SIP requirements that are associated with each classification.

TABLE 1.1.

Classification	Ozone Air Quality	Time Allowed (Years)	Major Source (Tons/Year)
Marginal	0.21–0.137 ppm	3	100
Moderate	0.138–0.159 ppm	6	50
Serious	0.160–0.179 ppm	9	50
Severe "1"	0.180–0.190 ppm	15	25
Severe "2"	0.191–0.279 ppm	17	25
Extreme*	0.280+ ppm	20	10

*Los Angeles.

A non-attainment area may fall from a classification higher on the chart to one that is lower if planned emission reductions do not achieve desired results. In such cases, the non-attainment area may receive more time to comply, but the requirements of its implementation plan are more extreme.

Following is a summary of the major plan provisions for ozone non-attainment areas.

1.4.5.1 Volatile Organic Compounds (VOCs)

Most non-attainment areas are required to reduce VOC emission by 3% per year to demonstrate "reasonable further progress" until the air quality standard is attained. The reduction is based upon a baseline emissions inventory (usually 1990), and both inventory and emission reduction amounts are generally expressed as tons of VOC emissions during a typical summer weekday. By using summer weekday emissions, air quality planners can adequately deal with emission categories that are highly seasonal, such as mobile source evaporative emissions, and can base their plans on data that represent conditions most likely to be experienced during high ozone levels.

1.4.5.2 Nitrogen Oxides

Consensus in the scientific community is that NO_x emissions exacerbate ozone non-attainment, even though they may scavenge ozone in some urban areas. Sources of NO_x may be required to install Available Control Technology (ACT) in non-attainment areas. ACT technology for NO_x reduction is generally based upon changes in burner design. Current ACT technology requirements are available through EPA and should be discussed with your local or state air quality planning officials.

1.4.5.3 Motor Vehicle Inspection and Maintenance (I/M)

I/M is a control measure that may be required for either carbon monoxide or ozone non-attainment areas. Experience with I/M in this country began in the early 1970s with operable programs in Chicago and New Jersey.

Provisions in the CLean Air Act discourage I/M programs that are conducted by private businesses where automobile parts and service are also provided. Any non-attainment area that has moderate or severe air quality and an urbanized population of 200,000 or more must implement I/M. If the area has ozone air quality that is serious or worse, it must implement an "enhanced" I/M program. The distinguishing features of such a program are

- collection of emissions over a simulated driving cycle
- canister purge system check
- evaporative control system check
- on-board diagnostic check
- minimum waiver repair cost of $450
- no waiver for tampered emission controls

Over thirty-five urbanized areas in the United States will be implementing enhanced I/M by federal requirement. However, several other urban areas may implement enhanced I/M for other SIP reasons. Many urban areas are having difficulty demonstrating that they can achieve the required 3% per year emission reduction requirement which is required by the Act in ozone non-attainment areas. Enhanced I/M remains to be one of the largest and most cost effective emission reduction measures available in many urban areas.

1.4.5.4 Gasoline Vapor Recovery

Ozone non-attainment areas will be required to implement gasoline vapor recovery at gasoline service stations. (In its simplest form, the vapor displaced from the automated gasoline tank is routed to the underground storage tank, which is under negative pressure.) This control measure has been successfully applied for years in the South-Coast Air Basin and in the Bay Area of California and is now required to be implemented in all serious and worse non-attainment areas.

1.4.5.5 Commercial and Consumer Products

After years of placing tighter controls on large, stationary sources and the automobile, emissions from commercial products have increased in both a relative and real sense. These products include mouthwash, paints, adhesives, household solvents, etc. These regulations will not be implemented by state law, but instead will be implemented by EPA rule, and applied to manufacturers, distributors, etc.

1.4.5.6 Carbon Monoxide

The Act establishes two classifications of carbon monoxide non-attainment — moderate (9.1–16.4 ppm) and serious (16.5 ppm and above). Moderate areas are required to comply by December 31, 1995; serious areas are required to comply by December 31, 2000. I/M is required in both areas, but enhanced I/M is required in severe areas. Carbon monoxide tends to be a localized problem, which is much more frequently associated with mobile sources than stationary fuel combustion or process sources.

1.4.5.7 Particulate Matter

The emphasis in the law has changed from "suspended" particulate (approximately 70 microns and smaller) to "respirable" particulates (10 microns and less in diameter) or PM_{10}.

Areas specified by EPA as "Group I" and any other areas that exceed the PM_{10} NAAQS before 1989 are designated non-attainment by law. These areas are required to meet the NAAQS by December 31, 1994. However, non-attainment areas that are designated after the initial designations under the Act are designated as either moderate or serious — moderate areas much comply within six years of designation; serious areas must comply by December 31, 2001, or ten years after designation, whichever is later. Requirements that may apply to serious areas can be waived if it is demonstrated that man-made sources do not significantly contribute to the problem. As with ozone, three-year incremental decreases are required; however, no annual or periodic goal is prescribed by law.

1.5 TITLE III: HAZARDOUS AIR POLLUTANTS (HAPs)

The tragedy of Bhopal, India, which lost 3400 people to a methyl isocyanate release, was on the minds of Congressmen when they considered the 1990 Clean Air Act Amendments. Other studies had shown that toxins in water may have an airborne genesis, which is also recognized in the Act. For example, benzene is emitted at gasoline pumps from unleaded gasoline at more than double the levels as compared to leaded gasoline. Benzene is suspected of being a carcinogen.

Because of the plethora of toxic substances, the CAA amendments are geared towards regulation of emission source categories, which constitute both areas and point sources.

1.5.1 MAXIMUM ACHIEVABLE CONTROL TECHNOLOGY (MACT)

A national goal has been established for a 75% reduction in HAP emission by the year 2000. The thrust of these regulations will be to determine what emission reductions are currently achievable in a full-scale plant and use it as a regulatory model. A major HAP source is one that emits 10 tons or more per year of a HAP, or 25 or more tons per year of a combination of HAPs. This level of control on new sources is referred to as "maximum available control technology" (MACT). MACT can also be applied to existing emission sources, which EPA has determined to be equal to the level of control achieved by the best 12% of existing sources, excluding some categories. The law also allows EPA to reevaluate the level of MACT based upon risk of cancer for a suspected compound. If a change is made, industry has eight years in which to adapt to the new standards.

Sources must comply with standards within three years of adoption, in most cases, with EPA being required to phase in standards for over a period of ten years. If EPA fails to adopt regulations for a source category, then the emission source can be certified to meet MACT by an independent engineer, if it is controlled within eighteen months of the time that the standard was to have been promulgated.

In addition to the three-year compliance term, if a source voluntarily agrees to reduce VOC emissions by 90% (or particulates by 95%), then it is eligible for an additional six years to comply.

1.5.2 RESIDUAL RISKS

Establishing actual risks from toxic materials, especially carcinogens, is not a clearly defined process. Congress has asked EPA and the Surgeon General to report to Congress by November 1996, their recommendations on a) how public health risks from toxic substances can be determined, b) the significance of these risks, and c) the technical/economic considerations in establishing standards of compliance.

1.5.3 ACCIDENTAL RELEASES

EPA must also develop a list of the "TOP 100" extremely dangerous substances.

1.5.4 WASTE INCINERATION

Standards are required to be adopted for hazardous waste incineration units, which, for regulatory purposes, are divided into three categories:

- small units (less than 250 T/day)
- large units (greater than 250 T/day)
- medical and infectious waste

Standards must be established for opacity, sulfur dioxide, hydrogen chloride, nitrogen oxide, carbon monoxide, lead, cadmium, mercury, dioxin and dibenzofurans. Standards are to be reviewed by EPA every five years.

For years, atmosphere release of toxic substances has been suspected of causing pollution of the Great Lakes and coastal waters. By November 1993, EPA must assess the impact of these emissions on water quality.

1.6 TITLE IV: ACID RAIN

Perhaps no other aspect of the Clean Air Act has been so politically sensitive as these provisions, since the areas that must bear the impact of emission controls (midwest power plants) are separated geographically and politically from the impacted areas (Northeast United States and Canada).

The solution was to adopt a plan that relies upon the marketplace to achieve an emission control program that is capable of achieving the emissions reduction goal at a minimum of economic burden.

Because of their tall stacks, emissions from utilities can be carried hundreds of miles before they are deposited in the form of halide salts or acid rain. The goal of the Act is to reduce 1990 emission approximately one-half by the year 2000. This goal is to be achieved through a system of trading emissions "allowances." Each allowance enables the holder to emit one ton of sulfur dioxide during a given calendar year. After the year 2000, any new or unplanned facility will be required to purchase allowances from other sources or the free market in order to operate.

The program is divided into two parts—Phase I begins with the passage of the Act and continues to December 31, 1994. During this time period, the market system is established and plans can be developed by the regulated community to prepare for Phase II, such as installing necessary monitoring equipment.

By law, EPA cannot allocate more than 8.9 million tons of SO_2 allowances after January 1, 2000. Each regulated emission source (110 facilities were referenced when the Act was passed) can only emit the emission quantities for which it was allowed or had subsequently purchased. Upon the implementation of Phase II, each source has several options for continued operation:

- purchase allowances on the market
- install control equipment
- implement conservation measures or reduced capacity

1.6.1 PHASE I REQUIREMENTS

EPA has identified 110 major utility plants (i.e., emission sources), located primarily in the midwest, that are covered by the Phase I program. These plants contain 265 generating units of 100 megawatts or more and emit SO_2 at a rate of 2.5 lbs/mm Btu or greater. Each of these units receives an emission allowance equal to 2.5 lbs/mm Btu multiplied by 40% of the unit's 1985–1987 average fuel consumption. If achieved, this reduction would achieve approximately 3.5 million tons of sulfur dioxide emission reduction by January 1, 1995.

Transferring Allowances—under Phase I, units may trade allowances among themselves or buy them at an auction from EPA. The purpose of the auction is to help establish an allowance market, or "prime the pump."

Allowances for construction of new facilities are reserved by EPA and are sold for up to $1,500 per ton. USEPA maintains an annual reserve of $75,000 of allowances for this purpose.

1.6.2 MONITORING AND RECORD KEEPING

Because of the "market value" of emissions, which are based on total emissions over the period of a year, EPA is requiring much tighter monitoring and record keeping of emissions. In-stack continuous monitors that sample for opacity, sulfur dioxide, nitrogen oxide and gas volume are required. In the case where more than one unit is vented by the same stack, separate continuous emission monitors (CEMs) are not required if sufficient operating data is available for each unit. A plant operator should take precautions to insure that the instruments to be used are of sufficient accuracy and reliability to meet the provisions of regulations and his own management requirements for quality data. Provisions for backup and data security should also be reviewed, since *all* data is necessary to accurately compare actual emissions with emission allowances. If the plant cannot support an emissions rate that compares favorably with the plant's allowances, significant penalties may be imposed. CEMs are required for plants subject to Phase I requirements by November 1994; Phase II compliance is required by December 1994.

1.6.3 PHASE II REQUIREMENTS

Beginning in the year 2000, the acid rain provision of the Clean Air Act will encompass all units with a generating capacity of 25 megawatts or larger. Those with 75 megawatts and larger capacity and emissions greater than 1.2 lbs/mm Btu will be eligible for a baseline emission equal to 1.2 lbs/mm Btu times the unit's baseline emission. This is expected to achieve approximately 65% reduction in emissions from these plants. The provisions of Phase I and Phase II apply, strictly speaking, to industrial boilers as well as utility units; however, it is expected that, because of size limitations, most industrial boilers impacted by the Clean Air Act Amendments will be impacted by Phase II, but not Phase I.

1.7 TITLE V: OPERATING PERMITS

The Act anticipates a nationally consistent—but state or locally administered—permit program to ensure that the emissions limitations under which a source is designed to operate are, in fact, consistently achieved in actual practice. The new requirements will be distinguished by much higher levels of monitoring, record keeping, and stiffer penalties for noncompliance. The spirit of the new permit requirements is accurately reflected in a sign once posted in an environmental engineer's office: "To err is human; to forgive is divine; neither of which is the policy of EPA!"

1.7.1 AFFECTED SOURCES

Final rules for Title V provisions were published July 21, 1992 (57 FR 32250) and are codified in Part 70 CFR Chapter 1, Title 40. They apply to all "major sources" and may also be applied to smaller sources at the discretion of the respective states. For most sources, the determination of whether or not they are "major sources" is based not on their emission levels, but rather on their potential to emit. For example, if a printing plant emits 50 tons per year in a marginal attainment area (100-ton-per-year major source cut-off limit), but only operates 2000 hours (i.e., one shift), its potential to emit may be over 200 tons per year if the plant were operating 24 hours per day, 365 days per year; therefore it would be a major source unless its construction or operating permit, issued under an EPA-approved permit program, conditioned the percent such that the emission from the source could never exceed 100 tons per year. For many sources that have actual emission limits that are much less than their potential to emit, permit conditions may allow the source to be classified as a minor source. Although this may not change the level of control that is required for the facility, it may eliminate the need for significant record keeping requirements and substantially decrease the plant's exposure to stiff federal penalties if real or perceived violations are later discovered by the state. This may only be a stop-gap measure, however, since EPA is required to study the feasibility of permitting non-major emission sources and may decide to do so.

The 1990 CAAA charged the EPA to promulgate emission standards for 40 categories and subcategories by November 15, 1992. This is one of many time schedules that has slipped. For example, currently MACT standards for utility boiler NO_x emissions will not be promulgated until April 30, 1995.

1.7.2 THE STATE ROLL

The Part 70 regulations will specify that local and state agencies must plan to implement the permit program. These plans, which constitute a part of each state's respective SIP, are due by November 15, 1993. In addition to procedural requirements, each state must demonstrate that they or a local permitting agency have

- an adequate enforcement program
- adequate staff
- an approvable fee structure
- a public information program

1.7.3 APPLICATION DEADLINE

Sources have until one year from the time that the permit program is approved in their state or local area to submit a permit application. The agency to which the application is sent has one year to act on the permit (§503).

1.7.4 INITIAL REVIEW

Anticipating that either companies filing for operating permits or citizens who disapprove the granting of a permit may want a quick review if such relief is not available from the state (presumably because of a backlog of work), any person may petition the state court to act on the permit or to require action by the state. If the federal EPA has not delegated authority to the respective state or local agency, then the relief must be sought in federal court (§502).

1.7.5 FEES

Although most states have charged nominal fees for reviewing permits in the past, the Clean Air Act Amendments require that states charge all of their costs to administrate the program to the permit applicants. In many states, this may pay for all air pollution control activities. Fees may not be less than $25/ton for all regulated pollutants, except CO. A state may charge a lower fee if it can demonstrate to EPA that such a fee will adequately cover state costs. A facility that fails to pay its fees may be charged a 50% penalty, plus interest (§502). In establishing a fee, the state is not required to count those emissions from any source that exceeds 4000 tons per year.

1.7.6 PERMIT SHIELD

Subject to public review and comment, a permit holder is shielded from prosecution under the Act for actions or circumstances that are otherwise contrary to the requirements of the Act but are in accordance with the terms and conditions of a granted permit. There are limitations to this shield, and some of its provisions may be resolved by EPA. Similarly, an applicant who has made a timely application for a permit will be afforded protection from the requirements to have a permit if the delay is not due to the source's failure to supply requested information. This protection is not available to sources subject to preconstruction review.

1.8 TITLE VI: ENFORCEMENT

Several mechanisms are available to EPA for enforcement of the Act against permit sources of pollutants. These include

- administrative penalties
- administrative orders
- civil actions by third parties
- criminal proceedings

1.8.1 PERMIT VIOLATIONS

EPA can seek injunctive relief and penalties for failure to comply with permit conditions or to violate a requirement of an approved state implementation plan. The source is advised of the alleged violation by receipt of a thirty-day notice of violation (NOV). Based on the notice, the facility can be penalized for violation either before or after the date of the notice. A facility should confer with legal counsel trained in environmental law before taking any action in response to an NOV.

1.8.2 ADMINISTRATIVE ORDERS

EPA has the authority to issue administrative orders that may become effective after the order recipient has had the opportunity to communicate with EPA on the facts of the matter at hand and the nature of the violation. Such orders do not take the place of penalties, nor are they allowed by law [§113(a)(4)].

1.8.3 CIVIL PENALTIES

EPA may assess civil penalties up to $200,000 (and higher, if the Attorney General concurs) for violations relative to most subjects in the Act that have been discussed in this chapter.

1.8.4 FIELD CITATION

Minor violations of the Act can be assessed through a "field citation program," where penalties of $5000 per day, up to $25,000 for a six-month period, may be assessed. USEPA considers the field citation program to be a warning: field citations do not preclude EPA from assessing additional penalties if the violations continue.

1.8.5 JUDICIAL REVIEW

Civil Penalty orders or field citations are subject to review if relief was sought within thirty days after the order became final. The review must be in U.S. District Court and can only involve material that constituted the record upon which the penalties were assessed [§707(1)(4)].

EPA has established written guidance, which it uses to determine the level of fines and penalties to be assessed. Several factors may be used to establish the fine, such as severity of pollution discharge, prior warnings, or penalties assessed, and any effort the facility takes to remediate emissions prior to a plant inspection. If an NOV is received, the environmental engineer should contact the required office to obtain a copy of the latest guidelines EPA is using to assess penalties. The document may be delivered, or you may call to obtain a copy through the Freedom of Information Process. It may take up to fifteen days to receive the information.

CHAPTER 2

Permitting Procedures

LYNDA M. WIESE

2.1 OVERVIEW

OFTEN, the maze of regulations surrounding the permitting and operation of air pollution sources can be as tortuous as the labyrinth. From a BACT (Best Available Control Technology) emission limit to acid rain credits, these requirements and how they are applied to sources define the conditions under which a source is legally allowed to operate. This chapter will review some of the existing regulations from a national perspective and look to several of the new requirements that will be undertaken as part of the Clean Air Act Amendments of 1990.

2.2 EMISSION LIMITS

Emission limits are set on categories of air pollution sources to control amounts of air pollution. States have set emission requirements on particulate matter, sulfur dioxide, nitrogen oxides, carbon monoxide, hydrocarbons, lead and other air emissions as part of their State Implementation Plans (SIP). Other emissions that may be limited are toxic air emissions, odors, fugitive dust and visible emissions. Sources may be required to meet emission limits even if they are not required to get a permit.

Many of the emissions limits are based on where the source is located. For instance, sources located in areas not meeting air quality standards may have more stringent emission limits to meet. Other limits may be based on type of fuel used. When looking at emission limits for a boiler, the source should keep in mind secondary emission sources. In the case of a coal-fired power plant, there may also be emission limits on the coal and ash handling for control of particulate matter.

17

The federal government has established source specific emission limits for several categories of boilers. They are presented in Table 2.1.

These New Source Performance Standards (NSPS) are adopted by EPA as state-of-the-art control on sources develops. It can be noted that the earlier NSPS require a certain emission rate to be met. Later proposals require a minimum control efficiency on the pollutants of concern.

In addition to the emission limits, these NSPS lay out the requirements for source performance testing, continuous emissions monitoring and periodic reporting.

2.3 CONSTRUCTION PERMITS—NEW SOURCE REVIEW

Many states have authority through their State Implementation Plans (SIP) to issue construction permits for new or modified sources. These permits contain the regulations that apply to that source. In addition, special requirements or conditions that may not be explicitly spelled out in the regulations would be contained in this document. A permit is a legally enforceable document. A source should be aware that the affected emission points and review times may vary from state to state. Most states also require that you receive your permit prior to start of construction. Check with the state or local permitting authority for the specific requirements in different areas.

Permit review usually begins with the submission of a permit application from the applicant or their consultant. If the source is anticipated to be large or controversial, it is recommended that the applicant meet with the permitting authority prior to submission of an application to determine what information is needed in the application.

The permitting agency will determine if the project needs a permit, and if so, if the application is complete. Requests for information may follow, asking the applicant to amend or complete their application.

The application review step insures that it is the permitting authority's belief that the project will meet any applicable emission limits and special requirements. Often, in lieu of having specific emissions data on a particular source, the agency will rely on EPA emission factors or source performance tests on similar equipment to determine potential emissions. The review must show that emission limits are being met and that the national ambient air quality standards are being met or maintained. A computer model is used to determine the effect of the source's emissions on the ambient air quality.

A source may be required to take more restrictive emission limits or operating restrictions to meet the regulations or ambient air quality standards. For fuel burning installations, this could take the form of restric-

TABLE 2.1. Emission Limits Since 1971.

Category	Citation	Effective Date	Requirements
Fossil fuel–fired steam generators	40 CFR Part 60 Subpart D	08/17/71	1.2 ppm Btu sulfur dioxide 0.10 ppm Btu particulate 0.2 to 0.8 ppm Btu nitrogen oxides (dependent on fuel used) continuous emissions and fuel monitoring
Electric utility steam generating units	40 CFR Part 60 Subpart Da	09/18/78	0.03 ppm Btu particulate % reduction for particulate % reduction for sulfur dioxide 0.2 to 0.8 ppm Btu nitrogen oxides depending on fuel used and % reduction Thirty-day average continuous emissions monitors and testing, periodic reporting
Industrial/commercial/ institutional steam generating units 100–250 million Btu per hour	40 CFR Part 60 Subpart Db	06/19/84	Sulfur dioxide emission limits and reduction requirements, nitrogen oxide and particulate emission limits, performance tests and continuous emissions monitoring
Boilers 10–100 million Btu per hour	40 CFR Part 60 Subpart Dc	09/12/90	Particulate limits for solid fuels, sulfur dioxide continuous emission monitoring if control device is being used, performance testing, no nitrogen oxide standard

tions on the amount of sulfur in fuels, height of stacks, or add-on control devices.

Companies should take a hard look at the preliminary proposal for permit issuance and insure that from their standpoint, the project can meet the requirements as proposed. A period for public comment follows with the opportunity for a public hearing on the agency proposal. USEPA may comment at this time also.

Additionally, the permitting agency will insure that the proposal complies with the National Environmental Policy Act (NEPA) and any state or local regulations on environmental assessments or environmental impact statements. States may charge fees to cover the processing of permits.

The new source review program directed by USEPA is the Prevention of Significant Deterioration (PSD) program. This permit program covers large new and modified sources of air pollution that want to locate in attainment areas (areas of the country that have demonstrated that they meet the national ambient air quality standards or are deemed unclassifiable). A major source that would be subject to the PSD program is any source that has the potential to emit more than 250 tons per year of any of the criteria pollutants. However, there is a list of twenty-eight named PSD categories (Table 2.2) that are considered major sources (and subject to PSD) if their criteria pollutant emissions are greater than 100 tons per year. Fossil fuel–fired steam electric generators and boilers are included in this table.

The significant emission levels for major modifications to major sources are included in Table 2.3. When emissions from a modification to a major source are anticipated to exceed this level, it triggers a PSD review. Sources subject to the PSD program undergo an intense review and approval period.

Emission limits set for PSD reviews must reflect the use of Best Available Control Technology (BACT). BACT is the maximum reduction in pollutants that is available at similar sources. The emission limits are set on a case-by-case basis and the aproach to setting them is to start with consideration of the best controls in place anywhere and discount the level of control due to technologic or economic infeasibility. This procedure is known as the top-down approach to setting a BACT standard.

BACT as applied to fuel burning installations could take the form of low-NO_x burners or selective non-catalytic or selective catalytic reduction (SCR) for nitrogen oxide control; wet or dry scrubbers for sulfur dioxide control; and baghouses for particulate matter and PM_{10} (particulate matter smaller than 10 microns) control. There may be other add-on controls or process or raw material changes that may be specified as BACT. The BACT/LAER clearinghouse, as maintained by USEPA is often used in determining the level of control to be evaluated.

The applicant for a PSD permit will also have to submit a detailed air

TABLE 2.2. Named PSD Source Categories.

Fossil fuel–fired steam electric plants of more than 250 million Btu per hour heat input
Coal cleaning plants (with thermal dryers)
Kraft pulp mills
Portland cement plants
Primary zinc smelters
Iron and steel plants
Primary aluminum ore reduction plants
Primary copper smelters
Municipal incinerators capable of charging more than 250 tons of refuse per day
Hydrofluoric acid plants
Sulfuric acid plants
Nitric acid plants
Petroleum refineries
Lime plants
Phosphate rock processing plants
Coke oven batteries
Sulfur recovery plants
Carbon black plants (furnace process)
Primary lead smelters
Fuel conversion plants
Fossil fuel boilers (or combinations thereof) totaling more than 250 million Btu per hour
  heat input
Petroleum storage and transfer units with a total storage capacity exceeding 300,000
  barrels
Taconite ore processing plants
Glass fiber processing plants
Charcoal production plants

quality analysis to show that the proposal will not endanger any air quality standards or increments. Once a complete application has been received for a major source, the ambient air quality baseline is set in the area affected by the source. Newer sources can only consume a portion of the leftover air quality or increment. PSD review requires that other impacts on the environment be examined and analyzed as part of the review. Public notices for permit reviews contain information on the amount of the ambient air quality increment that is consumed.

A PSD review may also require that preconstruction monitoring be performed by the source to determine the background emissions of a certain pollutant. This monitoring is normally continued into the operating phase for a period of time to insure that emissions from operation of the source, when added to the background emissions, do not exceed an air quality standard. Generally, this time period may not be less than one year.

Shifting now to major sources in non-attainment areas (areas in which one or more of the criteria pollutants is already determined to exceed the

TABLE 2.3. Significant Emission Rates [Source: 40 CFR 52.21(6)(23)].

Pollutant	Significance Level (Tons/Year)
Carbon monoxide	100
Nitrogen oxides	40
Sulfur dioxide	40
Particulate matter	25
PM_{10}	15
Ozone (VOC)	40 (of VOC's)
Lead	0.6
Asbestos	0.007
Beryllium	0.0004
Mercury	0.1
Vinyl chloride	1
Fluorides	3
Sulfuric acid mist	7
Hydrogen sulfide (H_2S)	10
Total reduced sulfur (including H_2S)	10
Reduced sulfur compounds (including H_2S)	10
Any other pollutant regulated under the Clean Air Act	Any emission rate
Each regulated pollutant	Emission rate that causes an air quality impact of 1 microgram per cubic meter or greater (24-hour basis) in any Class I area located within 10 km of the source

national ambient air quality standard). They have a similar program for permitting. Like PSD, affected sources receive a case-by-case emission limit that reflects the Lowest Achievable Emission Rate (LAER). Non-attainment area major sources are also required to secure a reduction in background emissions, called an "offset" greater than the amount of added emissions allowed under the permit. This is to assure the air quality in the non-attainment area is not worsened by the new or modified source.

One new development with the Clean Air Act Amendments of 1990 is the designation of new non-attainment areas. The new designations will carry with them lowered thresholds for sources that will be considered major and subject to non-attainment permit review in many areas. In some ozone non-attainment areas of the country, the sources requiring offsets and LAER would drop down to 25 tons of volatile organic compound or nitrogen oxide emissions. In Los Angeles, this threshold drops to 10 tons per year. The offsets required increase with the severity of the non-attainment designation. Small natural gas–fired boilers will be one of the sources required to comply with this regulation.

2.4 OPERATION PERMITS

Many state and local programs currently permit existing sources of air pollution. Through the Clean Air Act Amendments of 1990, there will be a federal operation permit program. These operation permits will have a maximum life of five years and must contain compliance schedules for sources that are currently not meeting their emission limits.

The regulation, Part 70 of the Code of Federal Regulations, pertains to major sources in both attainment and non-attainment areas. Operation permits require sources to comply with all the requirements in the Clean Air Act. Conversely, a source may be considered in compliance with the Clean Air Act if it is in compliance with its permit. This "permit shield" would shield sources from enforcement action on anything not restricted in the permit itself. The shield provision is optional (i.e., states are not *required* to adopt a permit shield) so the availability and scope of the shield may vary from state to state.

2.5 COMPLIANCE DEMONSTRATION

Once a permit is issued, or periodically based on state and local regu-ations, a source will be required to demonstrate that they operate in compliance. There are several ways of demonstrating compliance. Some regulations (such as the NSPS) list the preferred method of compliance demonstration. Other states have specific requirements for compliance demonstration in lieu of any federal requirements. The federal government will be adopting an enhanced monitoring policy for major sources to fol-low in demonstrating compliance.

A source emissions test or stack test is one method of compliance deter-mination. USEPA has set forth methods that sources or their consultants must use to perform these emissions tests. These methods are found in 40 CFR part 60 appendix A. Some sources are required to perform periodic emissions tests according to state or local requirements. Stack testing is not a practical alternative for continuous demonstration of compliance. A stack test represents a snapshot in time of the operation of a source. Operating variables recorded during a compliance stack test can be moni-tored more frequently to demonstrate compliance.

Other methods of compliance demonstration are fuel sampling and anal-ysis according to ASTM standards or installation and operation of con-tinuous emissions monitors. Sources may also be required to monitor op-erational parameters on control devices such as temperature, pressure, pressure differentials, drops or voltages to show the devices are functioning properly. Some of the NSPS require that continuous emission monitors (CEMs) be installed for opacity (visible emissions), sulfur dioxide, nitrogen oxides and either oxygen or carbon dioxide.

Sources are required to report periodically on the results of their compliance demonstration and follow performance guidelines for calibrating the equipment. In any case, the source is required to notify the agency whenever a standard is exceeded.

2.6 NEW REQUIREMENTS

With the passage of the Clean Air Act Amendments of 1990, EPA will be developing Maximum Available Control Technology (MACT) standards for toxic air emissions. On the initial list of proposed MACT categories (*Federal Register*, Vol 57, No. 157, 7/16/92) fuel combustion was listed. Of the 189 hazardous air contaminants that are proposed to be regulated by EPA, boilers may have standards for control of heavy metals to meet in the future.

Another Clean Air Act provision specifically affecting large coal-fired power plants is the acid rain control strategy in Title IV. There is a system of emission credits, assigned to sources, they can buy and sell to meet their annual emission limits set under the acid rain program. Sources that overcomply with the regulations can generate excess credits for selling on the open market. Trading of emission credits can be across the country. The idea behind this emissions trading program is to cap the emissions of acid rain precursors sulfur dioxide and nitrogen oxides on a national basis.

Use of other, non-traditional fuels is gaining momentum in many states. Industrial solid waste fuel pellets, municipal solid waste (refuse derived fuel) and wood waste are examples of alternate fuels that have been used in boilers. Each fuel has their own special characteristics. While reducing the amount of materials that goes into landfills, concern is usually raised over what the constituents are of the waste fuel and what emissions can be expected. Fuel burning installations are limited in the amount of these fuels they can fire before being considered an incinerator.

In summary, regulation of fuel burning installations, though varied, concentrates on emissions from the fuel(s) used. Sources need to be aware of emission limits and permit requirements for boilers and should check with the state/local agency personnel on how the regulations affect their specific boiler operation.

CHAPTER 3

Types of Boiler Systems and Combustion Fundamentals

FRANK L. CROSS, JR.
HOWARD E. HESKETH

3.1 INTRODUCTION

T HERE are a number of fuel systems that industry has used in the past for burning coal. The types that historically have been used are:

- hand stokers
- under-feed stokers
- traveling grate stokers
- chain grate stokers
- spreader stokers
- vibragrate stokers
- reciprocating grate stokers
- dump grate stokers
- cyclone furnaces
- pulverized coal

It is important to note that these types of coal systems and the size of these systems are quite different (see Table 3.1 and Figure 3.1). Table 3.1 lists area heat release based on ft² of grate area and volumetric heat release based on ft³ of burner volume.

Useful equivalencies applicable to coal-fired systems are listed in Table 3.2 along with some typical values for wood. Those are useful when actual data are not available.

3.2 AIR POLLUTION CONSIDERATIONS

The following factors affect dust loadings by the

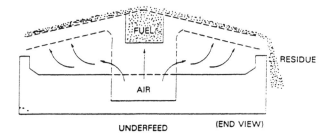

UNDERFEED   (END VIEW)

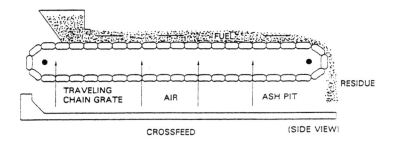

CROSSFEED   (SIDE VIEW)

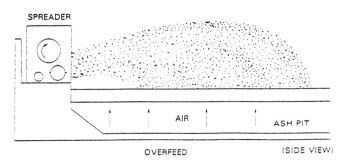

OVERFEED   (SIDE VIEW)

Figure 3.1.

TABLE 3.1. Stokers and Heat Release.

A. Types of Stokers
   Overfeed—spreader
              mass burning stoker
              (cross feed)
   Underfeed—single retort
              multiple retort
B. Heat Release Characteristics of Stokers

		Grate Heat Release Rate, Btu/hr·ft²	Furnace Heat Release Rate, Btu/hr·ft³
Single Retort			
Stationary Grate	(Conservative)	90,000	25,000
Moving Grate	(Conservative)	250,000	25,000
Stationary Grate	(Maximum)	125,000	32,000
Moving Grate	(Maximum)	350,000	32,000
Multiple Retort	(Conservative)	350,000	25,000
Multiple Retort	(Maximum)	450,000	35,000
Chain Grate	(Conservative)	375,000	25,000
Chain Grate	(Maximum)	475,000	32,000
Spreader Stoker			
Dump Grates	(Conservative)	375,000	25,000
Dump Grates	(Maximum)	400,000	32,000
Cont. Cleaning Traveling Grate	(Conservative	500,000	25,000
Cont. Cleaning Traveling Grate	(Maximum)	650,000	32,000

TABLE 3.2. Equivalencies.

1000 lbs/hr steam	10^6 Btu/hr
1000 lbs/hr steam	0.1 mW electrical generating capacity
1000 lbs/hr steam	0.04 tons/hr of 12,500 Btu coal
1000 lbs/hr steam	0.056 tons/hr of 9000 Btu wood
1000 lbs/hr steam	0.063 tons/hr of 8000 Btu wood
1000 lbs/hr steam	0.100 tons/hr of 5000 Btu refuse
1000 lbs/hr steam	200 scfm flue gases (70°F 1 atm, wet)
1 standard cord of wood	128 cubic foot pile = 4' × 4' × 8'

36.5 HP ≈ 1000 lb of wood at 4500 Btu/lb (45–50% H_2O) at 60% combustion efficiency

- fuel sizing
- size gradation of fuel as delivered to stoker hoppers
- ASM content
- fuel distribution to grate
- air distribution to grate
- grate heat release
- heat release per foot of grate width
- furnace heat release
- furnace configuration
- amount (%) of fly ash reinjection
- reinjection system design
- general operating condition of the boiler, such as setting integrity, excess air required, etc.

Other pollutants are usually estimated by using emission factors. EPA emission factors are given in Table 3.3.

These factors give a good comparison between the emissions of coal burning systems for various pollutants (see Table 3.3).

3.3 THE SPREADER STOKER

Various stokers (ten) have been listed in Section 3.1. However, the

TABLE 3.3. Uncontrolled Emission Factors for Anthracite Combustion.*

Boiler Type	Particulates**		Sulfur Oxides†		Nitrogen Oxides‡		Carbon Monoxide§	
	kg/Mg	lb/ton	kg/Mg	lb/ton	kg/Mg	lb/ton	kg/Mg	lb/ton
Pulverized coal fired			19.58	398	9	18		
Traveling grate stoker	4.6§§	9.1§§	19.58	398	5	10	0.3	0.6
Hand fed units	5	10	19.58	398	1.5	3		

*Factors are for uncontrolled emissions and should be applied to coal consumption as fired.
**Based on EPA Method 5 (front half catch).
†Based on the assumption that, as with bituminous coal combustion, most of the fuel sulfur is emitted as sulfur oxides. Most of these emissions are SO_2, with 0.1–0.3% SO_3. S indicates that the weight percent of sulfur in the oil should be multiplied by the value given.
‡For pulverized anthracite fired boilers and hand fed units, assumed to be similar to bituminous coal combustion.
§May increase by several orders of magnitude if a boiler is not properly operated or maintained.
§§Accounts for limited fallout that may occur in fallout chambers and stack breeching. Emission factors for individual boilers may range from 2.5–25 kg/Mg (5–50 lb/ton) and as high as 25 kg/Mg (50 lb/ton) during sootblowing.
Source: AP42 Emission Factors.

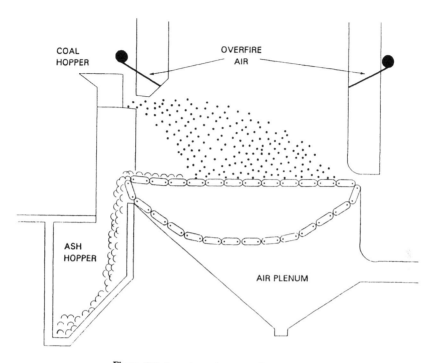

Figure 3.2 Spreader stoker – traveling grate type.

workhorse industrial type of coal burning system is the spreader stoker.

General limitations for the spreader stoker are 10,000 to 40,000 lbs. of steam per hour.

Note:

(1) Wide ability to either increase or decrease steam load rapidly
(2) Burned in suspension and the balance on the grates
(3) Grate designs:
   - lower capacities – dump grates
   - middle-sized units – continuous cleaning
   - higher capacities – traveling grates

The spreader stoker includes both suspension firing, along with the burning in a traveling grate (see Figure 3.2).

3.4 WASTE HEAT BOILERS

The heart of the energy recovery from the combustion flue gases is the boiler. This produces steam for use as live steam or for use in producing electricity. Boilers are basically heat exchangers. The type of boiler systems that have been limited to smaller systems are the waste heat boilers with fire tube systems (see Figure 3.3).

Waste heat recovery is a natural consideration when implementing a small boiler or an incineration system. The high-temperature gases typically generated offer an excellent source of energy with which to produce steam. Industrial plants generally have sufficient demand to fully utilize steam generated through a waste heat recovery operation.

For industrial or hospital applications, waste heat boilers of fire tube designs are most commonly used to affect the heat recovery process. A waste heat boiler is actually an ordinary fire tube boiler without a fuel burner. Exhaust gases are simply ducted from the fuel burner to the boiler to supply the energy source to produce steam. Like a traditional fire tube boiler, waste heat recovery boilers are available in one-, two-, three- and even four-pass designs.

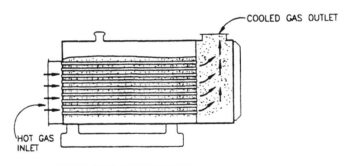

SINGLE PASS HEAT RECOVERY

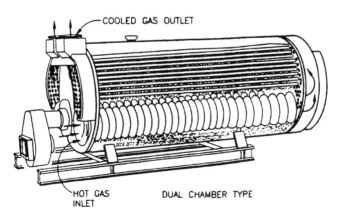

Figure 3.3 Waste heat boiler.

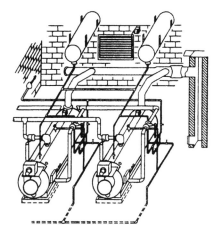

Figure 3.4 Boiler equipment and boiler room arrangement.

Basic operating procedures are essentially the same for a waste heat boiler as a traditional fire tube boiler. Proper feed water conditions must be maintained through chemical treatment. Solid content of boiler water must be controlled through blowdown.

There is one major concern, however, when operating a waste heat boiler with an incinerator. That is exhaust gas particulate content. Particulate, or fly ash, is a residue of the solid waste incineration process. Its presence simply means that a waste heat boiler must utilize a "dirtier" flue gas than a fuel-fired boiler. The primary effect of using a "dirty" flue gas is an increased fouling of the boiler tubes. Tube fouling hinders the heat transfer process and thus reduces the boiler efficiency. To alleviate this problem, soot blowers should be installed.

Other areas that may require periodic cleaning due to particulate buildup include the exhaust gas intake and the induced draft fan blades. Fouling of fan blades may decrease fan efficiency and thus increase the power required to maintain a proper draft on the system.

An excessive particulate concentration in the flue gas may indicate poor incinerator performance. Proper incinerator operation is therefore important in maintaining an efficient heat recovery process.

Accessories and auxiliary systems included with a waste heat recovery boiler will depend on the equipment available in the existing boiler plant. Existing feed water, chemical treatment, and blowdown systems may be used if appropriately sized to accommodate an additional boiler. Proximity to the existing main stream lines and boiler plant should be considered when sizing a waste-to-energy system (see Figure 3.4).

For certain applications, it may be advantageous to implement a multi-fuel–fired boiler, capable of using both waste heat and a traditional fuel to generate steam. A multi-fuel–fired boiler offers the flexibility of operating in three modes: 1) waste heat only, 2) natural gas (or other fuel) only, and 3) waste heat and a second fuel together to achieve the full rate capacity of the boiler. A significant advantage to this system is in eliminating cycling between a waste heat boiler and a separate traditional fuel-fired boiler. In some cases, several older existing boilers may be replaced by a new, higher efficiency, multi-fuel–fired boiler load.

3.5 SMALL BOILERS

3.5.1 SOOT BLOWING

Often, waste heat boilers include soot blower equipment to reduce foiling and improve thermal efficiency.

Effect of Fouling on Boiler Efficiency

Inches of Soot	Loss of Efficiency
1/32	9.5%
1/16	26%
1/8	45%
3/16	69%

Soot blowers should provide for pressurized steam or air for tube cleaning.

3.5.2 OPERATION AND MAINTENANCE OF SMALLER BOILERS

To improve thermal efficiency and achieve reduced air pollution emissions, regular operation and maintenance are essential. Proper operation and maintenance also reduce excess down time. A typical list of the operation and maintenance procedures can be seen in Table 3.4.

3.5.3 MASS AND ENERGY BALANCES

To determine the temperatures and gas flows from combustion systems needed for air pollution control systems, mass and energy balances must be calculated to size various types of equipment (i.e., boilers, APC, fans, stacks, etc.).

3.6 COMPARATIVE DATA FOR SYSTEM AIR POLLUTION CONTROLS WITH AND WITHOUT WASTE HEAT BOILERS

There are some major reasons why waste boilers should be added to incinerators including:

TABLE 3.4. General Operating and Maintenance Procedures for Small Boilers.

Description	Frequency	Comments
Operate soot blowers (if manual)	Daily	Excessive build-up may require more frequent operation.
Check flue gas temperature	Daily	A temperature increase may be an indication of poor heat transfer; boiler tubes may require more frequent cleaning.
Check flue gas pressure	Daily	An increase in pressure drop across the boiler may indicate fouling of boiler tubes.
Check steam pressure	Daily	Decrease in steam pressure may indicate overloading of boiler.
Check burner (for multi-fuel–fired boiler)	Daily	Periodic cleaning may be necessary to maintain optimum burner performance.
Check blowdown	Daily	Excessive blowdown may result in significant heat loss and thus reduce boiler efficiency.

(continued)

TABLE 3.4. (continued).

Description	Frequency	Comments
Check blowers	Weekly	Excessive fouling of fan blades may reduce fan efficiency and increase energy consumption.
Check gas inlet to boiler	Weekly	Excessive fouling of gas inlet may result in decrease of gas pressure
Check dampers	Weekly	Excessive fouling may adversely affect desired draft on system.
Check water treatment procedures	Weekly	Insufficient treatment may result in scale buildup and/or corrosion.
Check valves	Monthly	Valve stems, packing may leak, lubrication may be necessary. Clean and recondition as necessary.
Check motors	Monthly	Clean and recondition as necessary.
Check pumps	Monthly	Clean and recondition as necessary.
Check electrical systems	Monthly	Clean terminals and replace any defective parts.

TABLE 3.5. Comparison of Operating Parameters with and without Heat Recovery Boilers for Sound Air Pollution Control Systems.

Type System:	Venturi Packed Tower		Wet Scrubber		Dry Scrubber	
Heat Recovery:	w/HRB	w/o HRB	w/HRB	w/o HRB	w/HRB	w/o HRB
Water usage (gpm)	2.33	13.18	2.3	8.8	0.0	6.5
System flow (#/hr)	12,046	17,475	12,031	15,283	10,880	14,133
System pressure drop (inches water gauge)	35.0	35.0	12.0	16.0	12.0	16.5
Total HP req'd.	50.0	75.0	25.0	30.0	25.0	30.0
Equip. cost (dollars)	140,239	149,608	157,000	362,000	188,500	390,500

- less gas flow to control systems
- lower water rate for APC systems
- less system pressure drop
- equipment cost

See Table 3.5, where HRB is Heat Recovery Boiler, to compare operating parameters.

Examples of wet and dry scrubber control effectiveness are shown in Table 3.6. In general, wet absorbers are better for removal of gases and for condensing vapors. Dry scrubbers are better for particle removal. Dry scrubber arrangements are shown in Figure 3.5 for a wet/dry scrubber (a) and a dry/dry scrubber (b). These devices would follow the waste heat boiler if one is present. No cooling is achieved in the dry/dry scrubber as no liquid is introduced. The wet/dry scrubber slurry cools the gases and protects the filter bags, but care must be taken *not* to reach the dew point.

Table 3.7 shows generic data comparing same year costs for a wet and a dry scrubbing system in 1989 year dollars. (These data are for an incinerator type "boiler," but the comparison is valid.) Example costs of a filter baghouse control on a circulating fluidized bed (CFB) boiler are shown in Table 3.8 in 1985 year dollars.

TABLE 3.6. Typical Scrubber Performance.

	Wet Scrubber	Dry Scrubber
HCl removal	99%	95%
Particulate removal	50–85%	>99%
Trace metal removal	50–75%	>99%
Chlorinated toxic compounds (PCDD & PCDF)	<90%	>99%
Typical particulate emissions	0.03–0.10 gr/dscf @ 7% O_2	0.015 gr/dscf @ 7% O_2

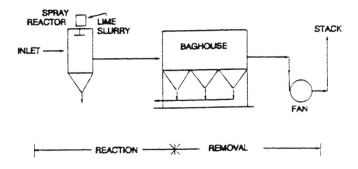

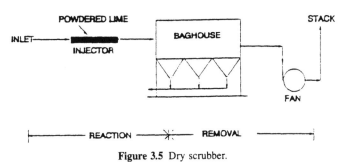

Figure 3.5 Dry scrubber.

TABLE 3.7. Typical Costs of Wet and Dry Scrubbers.*

	Wet Scrubber	Dry Scrubber
Capital Cost:		
APC unit installed	$733,500	$630,000
Annual Cost:		
Estimated annual operating time base 7200 hrs/yr (24 hr/day, 6 day/wk, 50 wk/yr)		
Electric	$107,130	$78,480
	(212.56 KW @ 0.07/KWII)	(155.24 KW @ 0.07 KWII)
Water	$2590	$1815
	(20 gpm @ $0.03/100 gal)	(14 gpm @ 0.03/100 gal)
Sewer	$1300	N/A
	(1.5 gpm @ $0.20/100 gal)	
Reagent	$43,200	$14,400
	NaOH	$Ca(O)_2$
	(30 lb/hr @ $400/ton)	(50 lb/hr @ $80/ton)
Fly ash disposal	N/A	$2,980
		(9.3 × yd³/yr @ $320/40yd³)
Total annual cost	$154,220	$97,670

*Application for 1900 lb/hr medical waste with 28,000 acfm flue gas at inlet temperature of 1800°F.

TABLE 3.8. Installation Summary—CFB Boiler with Pulse Jet Fabric Filter APC System.

Boiler — Circulating fluidized bed boiler
88,000 pounds per hour steam capacity
Fabric filter: 4 module pulse jet
Design fuel: Bituminous coal
16.7% Ash, 1.7% Sulfur

Specification	Design
Gas volume	34,285 acfm
Inlet temperature	340°F
Inlet grain loading	80 gr/acf
Outlet loading	0.035 lbs/10⁶ Btu
Pressure differential	6″ w.g.
Bag life	3 yrs.
No. of modules	4
No. of bags per module	156
Bag diameter	5.7″
Bag length	12′-0″
Gross air to cloth ratio	3.07:1
Net air to cloth ratio	4.09:1
Fabric material	Woven glass
Fabric weight	16 1/2 oz.
Fabric finish	10% Tef. B

3.7 COMBUSTION FUNDAMENTALS FOR FUELS WITH AIR

Combustion of fuels includes:

- evaporation of water (requires heat)
- volatilization of organics (requires heat)
- combustion of the fuel with air with the release of heat
- burner staging to assure adequate destruction of fuel components/ products by introduction of additional air and possibly using supplemental fuel

The supplemental fuels are normally either liquid or gas and use appropriate burners. Types of combustion are noted in Table 3.9.

The Flue Gas Products can be either as gas/vapor ($\uparrow$) or particles – note that gas/vapor can condense to liquid particles

Moisture $\rightarrow$ $H_2O\uparrow$

Solid and Liquid Combustibles:

Hydrogen $\rightarrow$ $H_2O\uparrow$

Oxygen – used in combustion

Nitrogen $\rightarrow$ $NO\uparrow$

C (including that in HC)

Mostly to $\rightarrow$ $CO_2\uparrow$

Little to $\rightarrow$ $CO\uparrow$

Very little to $\rightarrow$ C particles

Sulfur $\rightarrow$ $SO_2\uparrow$

Chlorine $\rightarrow$ $HCl\uparrow$

Fluorine $\rightarrow$ $HF\uparrow$

Ash ("non-combustibles" – These can be blown out with flue gas or left in bottom ash):

Asbestos (magnesium silicates) $\rightarrow$ particles

Mercury (from plastics, paper, metal, batteries, thermometers) $\rightarrow$ volatized as vapor$\uparrow$

TABLE 3.9. Types of Combustion.

Initial State		Fuel and Oxidant	
Fuel	Oxidant	Mixed	Separate
Gas	Air	Premixed flame	Diffusion flame
Liquid	Air	Premixed flame	Diffusion flame
Liquid	Liquid	Monopropellant combustion	—
Solid	Air	—	Diffusion flame
Solid	Solid	Propellant combustion, explosion	—

Lead } → Metal oxide vapors that condense easily to particle form
Antimony }

Beryllium → BeO particles

Arsenic, Cadmium
(from plastics, metal and batteries)
Chromium (from rubber, leather, metals) } metal oxide particles
Nickel
Ca, Na, Mg, Si, Fe, etc.

Additionally, there can be carryover to the flue gas of unburned or partially burned fuel as particles or vapors.

Currently, there are two proven methods for firing solid fuels in industrial boilers, stoker or pulverized fuel firing. The chronology of the fundamental combustion process as the fuel enters the furnace is basically the same for either firing system. In all systems, the "3 T's of combustion" are required, in addition to proper contacting of the fuel and the oxidant. These are: Time, Temperature, and Turbulence.

```
HEATING AND MINOR DEVOLATILIZATION
//////////////////////////////////////////→
          IGNITION
          /////////→
          MAJOR DEVOLATILIZATION
          ///////////////////→
          BURNING OF THE CARBON RESIDUE
          //////////////////////////////////////////////////////////////////→
0-------------------------------------------------------------------------------→
                   TIME AFTER ENTERING BOILER
```

Combustion is a chemical reaction and all reactions are reversible to some extent. Furthermore, no reaction ever goes to absolute completion. A stoichiometric combustion equation has just the amounts of reactants required by the balanced chemical equation for theoretical complete combustion. Complete combustion is when all carbon goes to CO_2 and all hydrogen goes to water. A reaction can be made more complete by increasing the concentration of the reactants; so in combustion operations, it is common to use more than the theoretical stoichiometric oxygen.

Stoichiometric combustion equation of a hydrocarbon in generalized equation form is

$$C_mH_n + \left(m + \frac{n}{4}\right)O_2 \rightarrow mCO_2 + \frac{n}{2}H_2O$$

Example:

$$\underbrace{CH_4}_{\text{fuel}} + (2)\left[\underbrace{O_2}_{\text{oxidant}} + \underbrace{\frac{79}{21}\,N_2}_{\text{diluent}}\right] \rightarrow \underbrace{CO_2 + 2H_2O}_{\text{comb. prod.}} + \underbrace{2\left(\frac{79}{21}\right)N_2}_{\text{diluent}}$$

(Air is considered to be 21% O_2 + 79% N_2)

Excess Air = Total Air − Theoretical Air

$$\%\,x\text{'s Air} = \frac{\text{Excess Air}}{\text{Theoretical Air}}\,(100)$$

Generalized Comb. Eq. w/Excess Air using CH_4 as example (a = moles excess air):

$$CH_4 + (2 + a)\left[O_2 + \frac{79}{21}\,N_2\right] \rightarrow CO_2 + 2H_2O$$

$$+ (2 + a)\frac{79}{21}\,N_2 + aO_2$$

Note: if a = 0.2 in this example, then

$$\%\,x\text{'s air} = \frac{0.2}{2.0}\,(100) = 10\%$$

3.8 CONVENIENT CORRELATION DATA

The desired end products from the combustion of fuels are CO_2 and H_2O as already noted. Other common products have also been listed. Flue gas concentrations of these materials in ppm can be converted to emission in lb/10⁶ Btu of heat input by mass and energy data. To simply this, Table 3.10 shows the normal correlation for some common fuels at 15% excess air.

TABLE 3.10. Typical Correlation of Units for Combustion Emissions at 15% Excess Air.

Emissions in lb/10⁶ Btu°F	Multiply Flue Gas Conc. in ppm by volume by:			
	#2 Fuel Oil	#6 Fuel Oil	Natural Gas	Propane
CO	7.76×10^{-4}	7.91×10^{-4}	7.30×10^{-4}	7.48×10^{-4}
VOC as CH_4	5.00×10^{-4}	5.00×10^{-4}	4.00×10^{-4}	4.00×10^{-4}
NO_x as NO_2	1.34×10^{-3}	1.34×10^{-3}	1.19×10^{-3}	1.23×10^{-3}
SO_2	1.86×10^{-3}	1.86×10^{-3}	NA	NA

NA = Not Applicable.

CHAPTER 4

Compliance Testing

FRANK L. CROSS, JR.

4.1 THE SOURCE TEST

4.1.1 INTRODUCTION

A source sampling experiment provides data on source emissions parameters. The isokinetic source test extracts a representative gas sample from a gas stream. Although often used only to determine compliance with emissions regulations, the test data can also provide information useful in evaluating control equipment efficiency or design, process economics or process control effectiveness. Valid source sampling experiments, therefore, yield valuable information to both the industrial and environmental engineer.

The source test is an original scientific experiment and should be organized and executed with the same care taken in performing any analytical experiment. This requires that objectives be decided before starting the experiment and that the procedures and equipment be designed to aid in reaching those objectives. The quantitative or qualitative analysis of the source sample should be incorporated as an integral part of the source test. After all work is done, the results should be evaluated to determine whether objectives have been accomplished. This section contains flow charts and descriptions to assist in the design, planning and performance of the source test described.

4.1.2 SOURCE TEST OBJECTIVES

The essential first step in all experiments is the statement of objectives. The source test measures a variety of stack gas variables that are used in

41

evaluating several characteristics of the emissions source. The source experiment should be developed with techniques and equipment specifically designed to give complete, valid data relating to these objectives. Approaching the experiment in this manner increases the possibilities of a representative sampling of the source parameters to be evaluated.

4.1.3 EXPERIMENT DESIGN

A well-designed experiment incorporates sampling equipment, techniques, and analysis into an integrated procedure to meet test objectives. The source sampling experiment must be based on a sampling technique that can collect the data required. The sampling equipment is then designed to facilitate the sampling procedure. The analysis of the sample taken must be an integral factor in the sampling techniques and equipment design. This approach of achieving test objectives provides the best possible source test program.

Designing a source test experiment requires a knowledge of sampling procedures and industrial processes, a thoroughly researched sampling experiment and a good basic understanding of the process operation to be tested. This knowledge assists in determining the types of pollutants emitted and test procedures and analyses that will achieve valid, reliable test results. A literature search of the sampling problem can yield information that may help improve test results or make testing much easier.

4.1.4 FINAL TEST PROTOCOL

The final test protocol clearly defines all aspects of the test program and incorporates the work done in research, experiment design and the presurvey. All aspects of this test, from objectives through analysis of the sample and results of the sampling, should be organized into a unified program. This program is then explained to industrial or regulatory personnel involved. This protocol for the entire test procedure should be understood and agreed upon prior to the start of the test. A well-organized test protocol saves time and prevents confusion as the work progresses.

4.1.5 TEST EQUIPMENT PREPARATIONS

The test equipment must be assembled and checked in advance; it should be calibrated following procedures recommended in the Code of Federal Regulations and this book. The entire sampling system should be assembled as intended for use during the sampling experiment. This assures proper operation of all the components and points out possible problems that may need special attention during the test. This procedure will assist in making preparations and planning for spare parts. The equipment should then be carefully packed for shipment to the sampling site.

The proper preparation of sampling train reagents is an important part of getting ready for the sampling experiment. The Method 5 sampling train requires well identified, precut, glass mat filters that have been desiccated to a constant weight. These tare weights must be recorded to ensure against errors. Each filter should be inspected for pinholes that could allow particles to pass through. The acetone (or other reagent) used to clean sampling equipment must be a low residue, high purity solvent, stored in glass containers. Silica gel desiccant should be dried at 250° to 300°F for 2 hours, then stored in airtight containers; be sure the indicator has not decomposed (turned black). It is a good procedure, and relatively inexpensive, to use glass-distilled, deionized water in the impingers. Any other needed reagents should be carefully prepared. All pertinent data on the reagents, tare weights, and volumes should be recorded and filed in the laboratory with duplicates for the sampling team leader.

4.1.6 TESTING AT THE SOURCE

The first step in performing the source test is establishing communication among all parties involved in the test program. The source sampling test team should notify the plant and regulatory agency of their arrival. All aspects of the plant operation and sampling experiment should be reviewed and understood by those involved. The proper plant operating parameters and sampling experiment procedures should be recorded in a test log for future reference. The sampling team is then ready to proceed to the sampling site.

The procedures for planning and performing the stack test for a basic Method 5 particulate sample are summarized in Table 4.1. Table 4.2 is a breakdown of the necessary source test events, and Figure 4.1 is a schematic of the Method 5 test train. Note that specific Method 5 stack sampling details are given in Chapter 12, Section 12.4.

4.2 TYPICAL STACK SAMPLING TEST PLAN

An abridged test plan is outlined in Table 4.3. This includes some sources not normally considered as emission sources. To this can be added others, especially the obvious emission control systems.

TABLE 4.1. **Planning and Performing a Stack Test.**

DETERMINE NECESSITY OF A
SOURCE TEST
• Decide on data required
• Determine that source test will give
  this data
• Analyze cost

STATE SOURCE TEST OBJECTIVES
• Process evaluation
• Process design data
• Regulatory compliance

DESIGN EXPERIMENT
• Develop sampling approach
• Select equipment to meet test
  objectives
• Select analytical method
• Evaluate possible errors or biases
  and correct sampling approach
• Determine manpower needed for test
• Determine time required for test with
  margin for breakdowns
• Throughly evaluate entire experi-
  ment with regard to applicable State
  and Federal guidelines

EACH STACK TEST
SHOULD BE
CONSIDERED AN
ORIGINAL
SCIENTIFIC
EXPERIMENT

RESEARCH
LITERATURE
• Basic process
  operation
• Type of pollutant
  emitted from
  process
• Physical state at
  source conditions
• Probable points of
  emission from
  process
• Read sampling
  reports from
  other processes
  sampled:
  1. Problems to
     expect
  2. Estimates of
     variables
     a. H_2O vapor
     b. Temperature
        at source
• Study analytical
  procedures used
  for processing test
  samples

PRE-SURVEY SAMPLING SITE
• Locate hotels and restaurants in area
• Contact plant personnel
• Inform plant personnel of testing
  objectives and requirements for
  completion
• Note shift changes
• Determine accessibility of sampling
  site
• Evaluate safety
• Determine past locations and ap-
  plication to Methods 1 and 2
• Locate electrical power supply to site
• Locate restrooms and food at plant
• Drawings, photographs, or blueprints
  of sampling site
• Evaluate applicability of sampling
  approach from experiment design
• Note any special equipment needed

FINALIZE TEST PLANS
• Incorporate presurvey into experi-
  ment design
• Submit experiment design for ap-
  proval by Industry and Regulatory
  Agency
• Set test dates and duration

TABLE 4.1. (continued).

CALIBRATE EQUIPMENT
- DGM
- Determine console ΔH@
- Nozzles
- Thermometers and thermocouples
- Pressure gages
- Orsat
- Pilot tube and probe
- Nomographs

PREPARE EQUIPMENT FOR TEST
- Assemble and confirm operation
- Prepare for shipping
- Include spare parts and reserve equipment

PREPARE FILTERS AND REAGENTS
- Mark filters with insoluble ink
- Desiccate to constant weight
- Record weights in, permanent laboratory file
- Copy file for on site record
- Measure deionized distilled H_2O for impingers
- Weigh silica gel
- Clean sample storage containers

CONFIRM TRAVEL AND SAMPLE TEAM ACCOMMODATIONS AT SITE

CONFIRM TEST DATE AND PROCESS OPERATION
- Final step before travel arriving at site

ARRIVAL AT SITE
- Notify plant and regulatory agency personnel
- Review test plan with all concerned
- Check weather forecasts
- Confirm process operation parameters in control room

SAMPLING FOR PARTICULATE EMISSIONS
- Carry equipment to sampling site
- Locate electrical connections
- Assemble equipment

DETERMINE APPROXIMATE MOLECULAR WEIGHT OF STACK GAS USING FYRITE AND NOMOGRAPHS

APPROXIMATE H_2O VAPOR CONTENT OF STACK GAS

PRELIMINARY GAS VELOCITY TRAVERSE
- Attach thermocouple or thermometer to pitot probe assembly
- Calculate sample points from guidelines outlined in Method 1 and 2 of *Federal Register*
- Mark pitot probe
- Traverse duct for velocity profile
- Record Δp's and temperature
- Record duct static pressure

RECORD ALL INFORMATION ON DATA SHEETS
- Sample case number
- Meter console number
- Probe length
- Barometric pressure
- Nozzle diameter
- C factor
- Assumed H_2O
- Team supervisor
- Observers present
- Train leak test rate
- General comments
- Initial DGM dial readings

USE NOMOGRAPH OR CALCULATOR TO SIZE NOZZLE AND DETERMINE C FACTOR
- Adjust for molecular weight and pitot tube C_p
- Set K pivot point on nomograph

LEAK TEST COMPLETELY ASSEMBLED SAMPLING TRAIN @ 15″ 11 g VACUUM AND MAXIMUM LEAK RATE OF 0.02 CFM

NOTIFY ALL CONCERNED THAT TEST IS ABOUT TO START

45

TABLE 4.1. **(continued).**

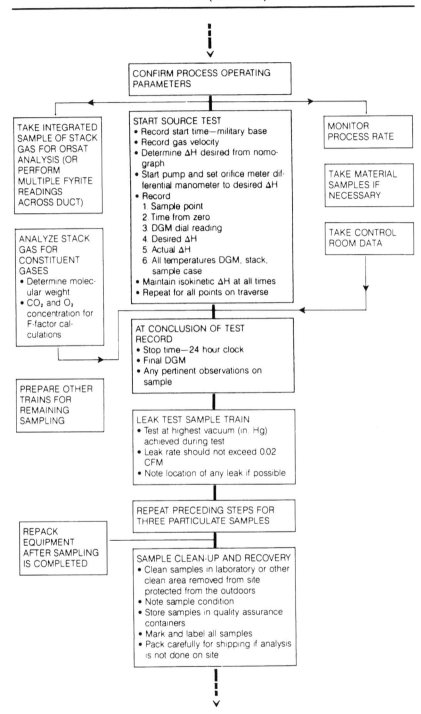

TABLE 4.1. **(continued).**

> **ANALYZE SAMPLES**
> - Follow *Federal Register* or State guidelines
> - Document procedures and any variations employed
> - Prepare analytical Report Data

> **CALCULATE**
> - Moisture content of stack gas
> - Molecular weight of gas
> - Volumes sampled at standard conditions
> - Concentration/standard volume
> - Control device efficiency
> - Volumetric flow rate of stack gas
> - Calculate pollutant mass rate

> **WRITE REPORT**
> - Prepare as possible legal document
> - Summarize results
> - Illustrate calculations
> - Give calculated results
> - Include all raw data (process of test)
> - Attach descriptions of testing and analytical methods
> - Signatures of analytical and test personnel

> **SEND REPORT WITHIN MAXIMUM TIME TO INTERESTED PARTIES**

4.3 EXAMPLE TEST PLAN

The following test plan developed for the Wisconsin DNR shows one sequence of logical events.

4.3.1 INTRODUCTION

(Company Name) has been retained by _____ to conduct compliance stack testing at the number two and number three foundry plants located in Waupaca, Wisconsin. The testing will be conducted in accordance with the Wisconsin Department of Natural Resources, code N.R. 439.07.

TABLE 4.2. Source Test Outline.

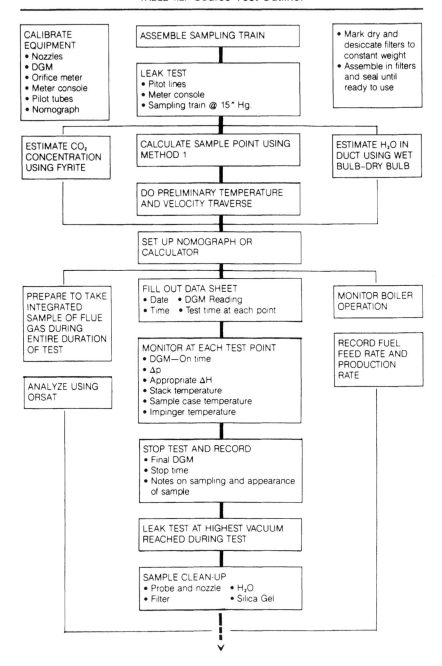

CALIBRATE EQUIPMENT • Nozzles • DGM • Orifice meter • Meter console • Pilot tubes • Nomograph	ASSEMBLE SAMPLING TRAIN	• Mark dry and desiccate filters to constant weight • Assemble in filters and seal until ready to use
	LEAK TEST • Pitot lines • Meter console • Sampling train @ 15″ Hg.	
ESTIMATE CO₂ CONCENTRATION USING FYRITE	CALCULATE SAMPLE POINT USING METHOD 1	ESTIMATE H₂O IN DUCT USING WET BULB-DRY BULB
	DO PRELIMINARY TEMPERATURE AND VELOCITY TRAVERSE	
	SET UP NOMOGRAPH OR CALCULATOR	
PREPARE TO TAKE INTEGRATED SAMPLE OF FLUE GAS DURING ENTIRE DURATION OF TEST	FILL OUT DATA SHEET • Date • DGM Reading • Time • Test time at each point	MONITOR BOILER OPERATION
ANALYZE USING ORSAT	MONITOR AT EACH TEST POINT • DGM—On time • Δp • Appropriate ΔH • Stack temperature • Sample case temperature • Impinger temperature	RECORD FUEL FEED RATE AND PRODUCTION RATE
	STOP TEST AND RECORD • Final DGM • Stop time • Notes on sampling and appearance of sample	
	LEAK TEST AT HIGHEST VACUUM REACHED DURING TEST	
	SAMPLE CLEAN-UP • Probe and nozzle • H₂O • Filter • Silica Gel	

TABLE 4.2. (continued).

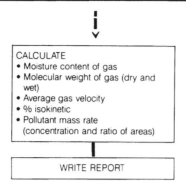

4.3.2 DETAILS OF THE TESTING

The testing will consist of EPA Modified Method 5 for condensible particulate matter with concurrent EPA Method 9 for opacity. EPA Method 10 utilizing integrated bag sampling will also be conducted for carbon monoxide (only the cupola scrubber). The testing will be conducted in accordance with the following Wisconsin Administrative Codes: NR415.05 for particulate matter, NR431.05 for visible emissions (opacity), and NR426 for carbon monoxide emissions. A complete synopsis of the meth-

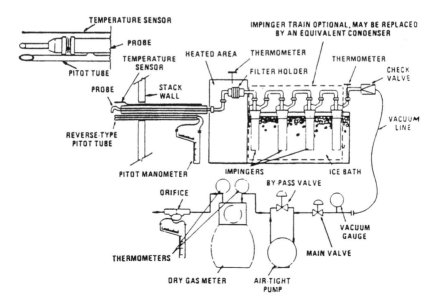

Figure 4.1 Method 5 particulate sampling train.

TABLE 4.3. Plant Testing Plan (Testing Date _____).

Source	Location	Inlet	Outlet	Parameter	Methodology
1) Pick 'n' Sort	No. 4 Plant	N/A	Yes	Preliminary Particulate	2—1 hr run utilizing EPA Method 5
2) Auto Pour	No. 4 Plant	N/A	Yes	Preliminary Particulate	2—1 hr run utilizing EPA Method 5
3) Millroom	No. 4 Plant	N/A	Yes	Preliminary Particulate	2—1 hr run utilizing EPA Method 5
4&5) Melt roof baghouse stacks	No. 4 Plant		Yes	Preliminary	2—1 hr run @ during
				Particulate	charging and w/o charging bricketter Velocity across all 7 fan outlets
6) Bricketter	No. 4 Plant	N/A	Yes	Preliminary Particulate	2—1 hr run utilizing EPA Method 5
	No. 4 Plant	N/A	Yes	Preliminary Formaldehyde	1—30 minute run utilizing Katz Method 122
7) Chip handling	No. 4 Plant	N/A	Yes	Preliminary Particulate	2—1 hr run utilizing EPA Method 5

odology to be utilized for the proposed testing is included in the attached appendices to this protocol.

4.3.3 SOURCE DATA

As illustrated in Table 1, DNR Testing Protocol, we have allowed a minimum of one full day for the testing of each permitted source. This schedule is tentative and subject to change due to inclement weather, production problems, or other adverse conditions beyond our control. Currently, we plan to commence daily at 8:00 A.M., starting the week of _____.

4.3.4 TESTING LOCATION CRITERIA

We have enclosed stack diagrams addressing each specific sampling location. The stack configuration on stacks Nos. (identify) are all identical. Stacks (identify) and the Cupola stack are as illustrated.

4.3.5 DESCRIPTION OF AIR FLOW

As illustrated in the following (identify) air flow diagrams:

Permit No. Casting Coolers P2; Sand System P5;
Molding Lines P9 (Plant No. 2)

Air from P-1 (cooling conveyors and shakeout Disa 1 vertical molding machine) and P-9 (Disa 2 vertical molding machine) is drawn to fan No. 1 inlet. Fan No. 1 exhaust air is released through (identify).

4.3.6 STACK TESTING LOCATION CRITERIA

Baghouse Stacks
(Number and identify)
Stack Dimensions
(Show for each)
Sampling Port Locations
3″ Sampling Ports are Located _____ (locate each Upstream and
_____ Downstream of any Obstructions
Traverse Points
(Give number and locate each)

4.4 MODIFIED METHOD 5 TEST METHOD FOR CONDENSIBLE PARTICULATE

4.4.1 PURPOSE

To describe a test method for particulate that includes the condensible particulate captured in the impingers of the United States Environmental Protection Agency Method 5 (EPA Method 5) as stated in Title 40, Code of Federal Regulations, Part 60, Appendix A or any other test method approved by the Department.

The test method shall be applied to all sources. The department may grant exemptions to New Source Performance Standard sources, sources to which Best Available Control Technology is applied, and other sources on a case-by-case basis.

4.4.2 SAMPLING EQUIPMENT

The sampling train and sample recovery equipment are the same as listed in EPA Method 5.

4.4.3 REAGENTS

The reagents are the same as listed in par. 3 of EPA Method 5 with the addition of methylene chloride-reagent grade. At the discretion of the agency and tester, any of the following reagent grade solvents may also be used: chloroform, ethyl ether, 1,1,1-trichloroethane, trichlorotrifluoroethane (freon 113) and any other solvent exhibiting similar properties as the solvents listed.

4.4.4 PROCEDURE

The sampling procedure is the same as listed in par. 4.1 of EPA Method 5. All parts of the sample collection portion of the train (nozzle, probe, cyclone and flask, filter holder, impinger glassware) shall be clean before sample collection. All sample collection portions of the train shall be made free of organic solvent residue before sample collection by rinsing them two or more times with distilled water.

The sample recovery procedure is the same as listed in par. 4.2 of EPA Method 5 with the addition of the following:

(1) The filter is placed in Container #1.

(2) The probe wash consisting of the probe rinse, the cyclone and flask

rinse, the connecting glassware rinse, and the front half of the filter holder rinse is placed in Container #2.

(3) The spent silica gel is placed in Container #3.

4.5 EXAMPLE OUTLINE OF A STACK TEST REPORT

1.0 Introduction
2.0 Process Description
3.0 Test Procedures
4.0 Summary of Test Results
5.0 Discussion and Conclusions
6.0 Particulate Testing Data (EPA Method 5)
    6.1 Particulate Testing Summary Sheet
    6.2 Particulate Sampling Analytical Procedures
    6.3 Stack Configurations
    6.4 Particulate Field Data Sheets
    6.5 Particulate Laboratory Work Sheets
    6.6 Particulate Calculations
    6.7 Particulate Equipment Calibration Data
7.0 Sulfur Dioxide Testing Data (EPA Method 6)
    7.1 SO_2 Testing Data Summary Sheet
    7.2 SO_2 Sampling Analytical Procedures
    7.3 SO_2 Field Data Sheets
    7.4 SO_2 Laboratory Work Sheets
    7.5 SO_2 Calculations
8.0 Sulfur Dioxide Testing Data (EPA Method 6C)
    8.1 SO_2 Testing Data Summary Sheet
    8.2 SO_2 Sampling Analytical Procedures
    8.3 SO_2 Calibration Data
9.0 Nitrogen Oxide Testing Data (EPA Method 7)
    9.1 NO_x Testing Data Summary Sheet
    9.2 NO_x Sampling Analytical Procedures
    9.3 NO_x Calculations
    9.4 NO_x Laboratory Data
    9.5 NO_x Flask Calibration Data
10.1 Nitrogen Oxide Testing Data (EPA Method 7E)
    10.1 NO_x Testing Data Summary Sheet
    10.2 NO_x Sampling Analytical Procedures
    10.3 NO_x Calibration Data
11.0 Visible Emissions Testing Data (FDER Method 9)

TABLE 4.4. **Summary of EPA Test Methods.**

Method		Reference			Description
1–8		42 FR 41754	08/18/77		Velocity, Orsat, PM, SO_2, NO_x, etc.
		43 FR 11984	03/23/78		Corr. and amend. to M-1 thru 8
1–24	C	52 FR 34639	09/14/87		Technical corrections
		52 FR 42061	11/02/87		Corrections
1		48 FR 45034	09/30/83		Reduction of number of traverse points
1	R	51 FR 20286	06/04/86		Alternative procedure for site selection
1A		48 FR 48955	10/21/83	P	Traverse points in small ducts
2A		48 FR 37592	08/18/83		Flow rate in small ducts—vol. meters
2B		48 FR 37594	08/18/83		Flow rate—stoichiometry
2C		48 FR 48956	10/21/83	P	Flow rate in small ducts—std. pitot
2D		48 FR 48957	10/21/83	P	Flow rate in small ducts—rate meters
3A		51 FR 21164	06/11/86		Instrumental method for O_2 and CO_2
3R		48 FR 49458	10/25/83		Addition of QA/QC
4	R	48 FR 55670	12/14/83		Addition of QA/QC
5	R	48 FR 55670	12/14/83		Addition of QA/QC
5	R	45 FR 66752	10/07/80		Filter specification change
5	R	48 FR 39010	08/26/83		DGM revision
5	R	50 FR 01164	01/09/85		Incorp. DGM and probe cal. procedures
5	R	52 FR 09657	03/26/87		Use of critical orifices as cal stds
		52 FR 22888	06/16/87		Corrections
5A		47 FR 34137	08/06/82		PM from asphalt roofing (P as M-26)
5A	R	51 FR 32454	09/12/86		Addition of QA/QC
5B		51 FR 42839	11/26/86		Nonsulfuric acid particulate matter
5C		Tentative			PM from small ducts
5D		49 FR 43847	10/31/84		PM from baghouses
5D	R	51 FR 32454	09/12/86		Addition of QA/QC
5E		50 FR 07701	02/25/85		PM from fiberglass plants
5F		51 FR 42839	11/26/86		PM from FCCU
5F	R	53 FR 29681	08/08/88		Barium titration procedure
5G		53 FR 05860	02/26/88		PM from woodstove—dilution tunnel
5H		53 FR 05860	02/26/88		PM from woodstove—stack
6	R	49 FR 26522	06/27/84		Addition of QA/QC
6	R	48 FR 39010	08/26/83		DGM revision
6	R	52 FR 41423	10/28/87		Use of critical orifices of FR/Vol meas.
6A		47 FR 54073	12/01/82		SO_2/CO_2
6B		47 FR 54073	12/01/82		Auto SO_2/CO_2
6A/B	R	49 FR 09684	03/14/84		Incorp. coll. test changes
6A/B	R	51 FR 32454	09/12/86		Addition of QA/QC
6C		51 FR 21164	06/11/86		Instrumental method for SO_2
		52 FR 18797	05/27/87		Corrections
7	R	49 FR 26522	06/27/84		Addition of QA/QC
7A		48 FR 55072	12/08/83		Ion chromatograph NO_x analysis
7A	R	53 FR 20139	06/02/88		ANPRM
7A	R	Tentative			Revisions

4.6 STANDARD AND SPECIALIZED TESTING

In many situations, special testing methods and analytical procedures are required. Table 4.4 summarizes EPA standard and special test methods. Once samples have been taken, analyses can be performed. Some of the various analyses that may be needed are listed in Table 4.5. Special sampling trains may also be required for low level concentration

TABLE 4.5. **Example Analytical Procedures.**

Sample	Analysis Parameter	Sample Preparation Method	Sample Analysis Method
1. Liquid waste feed	V-POHCs	8240	8240
	SV-POHCs	8270	8270
	Cl⁻	—	E442-74
	Ash	—	D482
	HHV	—	D240
	Viscosity	—	A005
2. Solid waste feed	V-POHCs	8240	8240
	SV-POHCs	8270	8270
	Cl⁻	—	D-2361-66 (1978)
	Ash	—	D-3174-73 (1979)
	HHV	—	D-2015-77 (1978)
3. Ash	V-POHCs	8240	A101
	SV-POHCs	P024b, P031	A121
	Toxicity	—	C004
4. Stack gas			
a. MM5 train			
Filter and	Particulate	M5	M5
probe rinse	SV-POHCs	P024b, P031	A121
Condensate	Cl⁻	—	325.2
	SV-POHCs	P021A	A121
XAD resin	SV-POHCs	P021A	A121
Caustic impinger	Cl⁻	—	325.2
b. VOST	V-POHCs	—	A101
c. Tedlar gas bag	V-POHCs	—	A101*
d. Gas bag	CO₂, O₂	—	M3 (Orsat)
e. Cont. monitor	CO	—	Continuous monitor

*Tedlar gas bag samples will be analyzed for V-POHCs, only if VOST samples are saturated and not quantifiable.
Note:
Four-digit numbers denote methods found in "Test Methods for Evaluating Solid Waste," SW-846.
Numbers with prefixes of A, C, and P denote methods found in "Sampling and Analysis Methods for Hazardous Waste Combustion."
Method No. 325.2 (for Cl⁻) is from "Methods for Chemical Analysis of Water and Wastes," EPA-600/4-79-020, March 1979.
Numbers with prefixes D and E denote methods established by the American Society for Testing and Materials Standards (ASTM).
M3, M5 refer to EPA testing methods found in the *Federal Register*, Vol. 42, No. 160, Thursday, August 18, 1977.

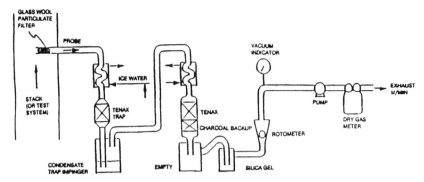

Figure 4.2 Volatile organic sampling train (VOST).

and the special samples needed. The schematic for one such train, the VOST, is in Figure 4.2.

An example of personnel required is noted in Table 4.6. Often, nine persons are required for each test, and often, only one test/day can be run. Triplicate tests are required. A typical listing of data forms is given in Table 4.7. Units of measure must be shown for each item in Table 4.7.

TABLE 4.6. A Typical Example of Sampling Personnel Required.

Job	Number of Personnel	Experience Required
1. Sample liquid feed (ounce every 15 min)	1	Technician with sampling experience and safety training
2. Drum solid sampling and recording (once every 5–10 min)	1	Technician with sampling experience and safety training
3. Sampling ash and scrubber waters every 1/2–1 hr	1	Technician with safety training
4. Stack sampling MM5	2	Experienced console operator and technician for probe pushing
VOST	1	Experience with VOST operation
5. Process monitor to record operating data every 1/4–1/2 hr and determine waste feed rates	1	Engineer or other person experienced in plant operations and trial burn requirements
6. Field laboratory	1	Experienced chemist for check-in and recovery of all samples, and preparation of sampling equipment for each run
7. Crew chief	1	Person experienced in all aspects of trial burn sampling to direct all activity and solve problems that may occur
Total	9	

TABLE 4.7. **Example List of Data Forms.**

Traverse Point Locations
Preliminary Velocity Traverse
Method 5 Data Sheets
Isokinetic Performance Work Sheet
M5 Sample Recovery Data
Integrated Gas Sampling Data (Bag)
Orsat Data Sheet
VOST Sampling Data

Drum Weighing Record
Drum Sampling Record
Liquid Waste Feed Sampling Record
Fuel Oil Sampling Record
Drum Feed Record

Process Data (Control Room)
Miscellaneous Process Data (In-Plant)
Tank Level Readings

Log of Activities

Ash Sampling Record
Scrubber Waters Sampling Record
Sample Traceability Sheets
GC/MS Data Calculation Sheets

Note: Units of measure must be shown for each item on each data sheet.

4.7 CONTINUOUS MONITORS

4.7.1 OPACITY MONITORING

4.7.1.1 Introduction

Opacity is a measure of degree to which the emissions from a stack obscure the view of an object in the background. Opacity readings indicate the visibility obscuring properties of the total gas stream, not the particulates alone. Opacity monitoring does not provide readings of particulate concentration. If the exhaust stream is void of water droplets, then the correlation between opacity and particulate concentration has been, in some instances, demonstrated to be quite accurate.

Transmissometers are used to measure opacity and operate by passing a beam of light through the exhaust stack. Part of the light is scattered and absorbed by particulate matter in the flue gases, while the remaining light reaches the transmissometer's detector. By comparing the amount of light transmitted to the amount recieved, the transmissometer calculates the opacity of the plume. A typical double pass transmissometer system is shown in Figure 4.3.

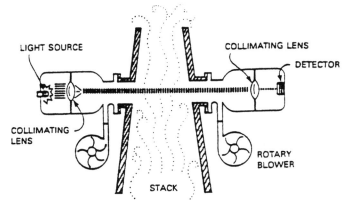

SINGLE-PASS OPACITY MONITOR

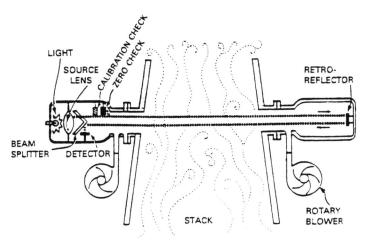

DOUBLE-PASS OPACITY MONITOR

Figure 4.3.

Double pass units have a transmitter and receiver on one side of the stack and a mirror on the other, while a single pass unit has the transmitter and receiver mounted on opposite sides of the stack. A double pass unit has greater sensitivity and, generally, has less operational problems.

Title 40, Part 60 Appendix B of the Code of Federal Regulations details the EPA requirements concerning opacity continuous emission monitoring. These regulations include installation specifications, design specifications, verification procedure, and performance specification procedure. The state of Arizona has adopted these regulations in the Arizona test manual.

4.7.1.2 Installation Concerns

Installation concerns include plume stratification, vibration, and ease of access. Placing the transmissometer where stratification (layering of gases) occurs is to be avoided since a non-representative measurement will be obtained. Locations to be avoided include horizontal ducts and ducts with sharp bends. The Federal Regulations help avoid stratification problems by specifying possible installation locations. Placement near a vibration source, such as induced draft fan, can cause optical alignment problems and is to be avoided. The transmissometer should be placed where personnel can easily access the unit for maintenance.

4.7.1.3 Operation Problems

Operational concerns include optical window fouling and calibration. To avoid erroneous readings, the optical windows must be kept clean. Most transmissometers include some type of air purging system or blower to aid in this endeavor. Even with this system, periodic removal and cleaning of the optical windows is necessary. An EPA approved unit (one meeting the previously cited Federal Regulations) allows for the windows to be cleaned without removing the unit or recalibrating it. As with almost any measuring instrument, calibration is a concern. The double pass transmissometers expedite the process by mounting a zeroing mirror and calibration filter in the transmitter/receiver unit. The zeroing mirror reflects the source light back through the clean interior of the unit. This simulates a clean stack condition and allows for calibration of the zero value. The calibration filter can then be moved into the light path to provide an upscale calibration value.

4.7.1.4 Transmissometer Economics

Equipment prices and associated startup costs for double pass transmissometers are shown in Table 4.8. As seen in Table 4.6, a fully op-

TABLE 4.8. Double Pass Transmissometer Costs (in 1988 Year Dollars).

Company	Model	Equipment Cost ($)	Startup Cost ($)	Data System Cost ($)	Total Initial Cost ($)
Datatest	90AS	8295*	10,000–12,000	2500	20,795–22,795
	900RM	14,485	10,000–12,000	2500	26,985–28,985
Leag-Siegler	1100m	15,650	8000–11,000	2500	26,150–29,150

*Customer must supply microprocessor.

erational EPA certified transmissometer will be in the $20,000–$30,000 price range. Estimates for a one-year service contract varied from $3000–$12,000.

4.8 COMBUSTION GAS MONITORING

4.8.1 INTRODUCTION

Carbon monoxide (CO) or oxygen (O_2) monitoring indicates the amount of CO or O_2 present in the exhaust stream. There is a variety of monitoring techniques that can be employed for either gas. Unlike opacity and its varying correlation with particulate concentration, all CO and O_2 monitoring techniques afford a direct correlation between the measured value and the actual concentration. This type of measurement produces reliable results that can be used with confidence.

Continuous emission monitors for gases are often categorized as in situ or extractive. With in situ equipment, gas analysis takes place inside the stack, while extractive systems remove a sample and condition it before analyzing it. Both CO and O_2 can be measured by in situ or extractive methods.

4.8.2 CO Monitoring

CO monitoring, whether in situ or extractive, utilizes infrared absorption spectroscopy to measure the CO concentration in the flue gases. Infrared energy is transmitted through the exhaust where portions of it are absorbed and the remainder reaches the detector where it is translated into a CO reading. With in situ system the infrared source and detector are mounted onto the exhaust stack and the infrared beam is transmitted directly through the stack. Another type of system uses a probe to remove an exhaust sample and send it to an infrared analyzer located in close proximity.

The extractive CO system employs a probe to remove a sample and a pump is then used to transmit the sample to a remote analyzer. Before reaching the analyzer, the sample is screened to remove particulate, moisture is condensed and removed, and gas temperature is lowered. The devices that perform these operations are collectively referred to as the handling or treatment system. Once the sample reaches the analyzer, the same infrared principles are used.

Federal regulations set forth by the EPA are found in the same section as those for transmissometers, that is Title 40, Part 60 Appendix B. These regulations for CO continuous monitors are not as encompassing as those for transmissometers; in particular, there are no design requirements.

Areas of concern with in situ monitoring include particulate, moisture, vibration and path length. Particulate matter interferes with the infrared beam and leads to inaccurate readings. Moisture in the form of water droplets tends to scatter the infrared beam while water vapor tends to absorb it; both introduce errors into the measurement process. In general, wet or dirty streams are not suitable for in situ devices. Sampling path length is a consideration for units that transmit through the exhaust stream. When the path length becomes too short, the monitor may not be sensitive enough to produce accurate readings. Vibration can be a problem for microphone type detectors and electronic and mechanical dampening may be required.

Extractive system analyzers are ideally located in an environmentally controlled room. High ambient temperatures can cause an analyzer located in an uncontrolled room to malfunction. Extractive systems also can have considerable lag times, if the distance between the probe and analyzer is excessive. For example: A sample velocity estimated at 3 feet/second means a sample traveling 180 feet would take one minute to reach the analyzer. This lag is not a particular concern when monitoring to show compliance. When monitoring with regard to combustion control, the lag can prove to be a hindrance. This is because the effects of altering the combustion parameters cannot be immediately evaluated.

Equipment and startup costs for CO in situ and extractive systems are shown in Table 4.9.

TABLE 4.9. CO Monitoring System Costs (in 1988 Year Dollars).

Company	Type	Model	Equipment Cost ($)	Startup Cost ($)	Data System Cost ($)	Total Initial Cost ($)
Rosemount Analytical	In situ (CO)	5100	12,000	5,000	2000	19,000
Datatest	Extractive (CO and O₂)	400	27,300	20,000	2250	29,550

4.8.3 O_2 MONITORING

In situ O_2 continuous monitoring is most often done using zirconium oxide techniques. A probe containing a solid electrolyte sensing cell samples the gas stream. The cell is constructed of stabilized zirconium oxide with platinum electrodes on the inner and outer surfaces. When the two surfaces are exposed to different oxygen concentrations, a potential is generated across the electrodes. One surface is exposed to a reference concentration of oxygen, so the potential generated is related directly to the O_2 concentration of the flue gases.

Extractive O_2 continuous monitors can employ zirconium oxide technology or a paramagnetic principle. Before reaching the analyzer, the sample is conditioned by a handling system.

The handling system is the same as that previously described for the CO extractive system. In fact, one handling system could serve CO and O_2 analyzers. Paramagnetism can be thought of as a material's ability to become a temporary magnet when placed in a magnetic field. O_2 is strongly paramagnetic and most other gases are not. This unique characteristic of O_2 can be used to measure O_2 concentrations.

Federal regulations promulgated by the EPA are found in the previously cited Title 40, Part 60 Appendix B. They are similar in scope to those for CO and like the CO regulations, no design requirements are included.

In situ O_2 monitors have filters to remove particulate, but the filters are not foolproof. Flue gases heavily laden with particulate are likely to cause operational problems.

An in situ device or extractive probe will require more attention when exposed to corrosive gases. This should not be overlooked in light of the high chlorine content of most hospital waste streams. The chlorine that goes into the incinerator in the form of plastics will come out of the stack as hydrochloric acid—a corrosive gas that will reduce the life span of any monitoring device.

As previously stated, extractive system gas analyzers can malfunction when exposed to high ambient temperatures. This concern applies equally

TABLE 4.10. O_2 Monitoring System Costs (in 1988 Year Dollars).

Company	Type	Model	Equipment Cost ($)	Startup Cost ($)	Data System Cost ($)	Total Initial Cost ($)
Rosemount Analytical	In situ (CO)	5200	3250	4000	2000	9250
Datatest	Extractive	400	27,300	20,000	2250	49,550

to CO and O_2 analyzers. The earlier discussion concerning the lag time of extractive systems also applies to either CO or O_2 systems.

The equipment and startup costs of both in situ and extractive O_2 monitoring systems are shown in Table 4.10. The Datatest figures include CO and O_2 monitoring systems. If Rosemount Analytical were to supply in situ CO and O_2 systems, the total cost would be approximately $28,250. It can be concluded that the extractive system would be between 1.5 and 2.0 times the cost of a comparable in situ system.

CHAPTER 5

Survey of Air Pollution Control Techniques

FRANK L. CROSS, JR.

5.1 INTRODUCTION

THE systems that require control for particular and acid gases (i.e., HCl/SO$_2$) for the smaller industrial plants (boilers, hospital incinerators, hazardous waste incinerators, etc.) include multiple processes. Generally, these systems include:

(1) *Particulates* (including possible ash, metals, etc.)
  - cyclones (mechanical collectors—used for high loadings with precleaners such as stoker coal burning systems)
  - Venturi type scrubbers (i.e., Anderson, Calvert, ACI)
  - baghouses
(2) *Acid gases* (SO$_2$ and HCl)
  - packed towers
  - dry scrubbers
  - other type absorbers

A combination of devices may be included in the air pollution control system as discussed in Chapter 3. Also, there may be precooling prior to the air pollution device(s) (see Figure 5.1).

5.2 DATA REQUIRED FOR AIR POLLUTION CONTROL SYSTEMS

Information data for the carrier gas, emissions parameters, and special data (i.e., particle size, resistivity, etc.) is necessary to determine the type and size of control device required. These are listed in Table 5.1.

65

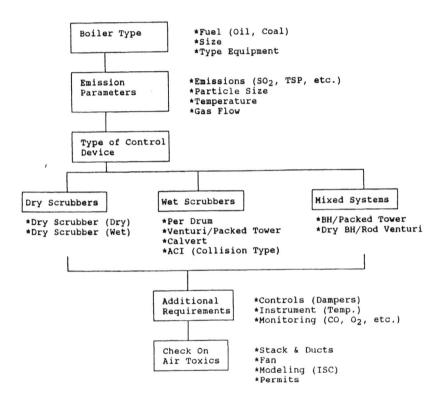

Figure 5.1 Retrofit for industrial furnaces for direct pollution control.

TABLE 5.1. **Air Pollution Control Device Specification Parameter Checklist.**

I. Carrier Gas:
   A. Airflow: _____ cfm @ _____ °F, _____ ″ Hg and _____% moisture by volume
   B. Composition: _____% carbon dioxide, _____% oxygen, _____% nitrogen
   C. Density: _____ lb/ft³ _____ °F, _____ ″ Hg
   D. Viscosity: _____ centipoise

II. Particulate Emission Parameters:
   A. Concentration: _____ grains/cubic foot @ _____ °F, _____ ″ Hg pressure, and
      _____% moisture by volume
   B. Mass Rate: _____ lb/hr
   C. Size Distribution: _____% less than 5 micrometers, _____% less than 15
      microns, _____% less than 25 microns (by weight) collected with a _____
      _____ (brand name), _____ (types
      of particle sizing equipment such as Bacho, Cascade impactors, sieve analysis,
      microscope), _____ (in situ or extractive) and reported on a _____
      _____ (weight or count) basis.
   D. Completeness of Combustion: _____% ash at a constant weight to _____ °F of
      sample collected on _____ (description of how
      and where sample was collected).
   E. Particle Characteristics: _____ (lb/ft³) bulk density (with voids), _____ (lb/ft³) true
      density (without voids), _____ (is or is not) wettable and/or _____
      (will or will not) go into slurry with water within 30 seconds contact time, _____,
      _____ pH (check at two concentrations resulting when thoroughly mixing with
      ample quantity of distilled-deionized (DDI) water @ a pH = 7, _____ (lb/lb) fil-
      trate to total particulate entering DDI solution phase (i.e., measure of disposable
      sludge), _____ (granular or fibrous) in character, _____
      (strong or weak) potential for the particle to cause mechanical abrasion at high
      velocity points and _____ & _____ chemical formula and concentration when
      absorbed in _____ (volume) DDI water.
   F. Resistivity: _____, _____, _____ ohm-centimeter resistivity @ _____, _____,
      _____ °F, _____, _____, _____% moisture and _____, _____, _____ ppm
      sulfur trioxide, respectively, measured _____ (in situ or extractive) with
      _____ (brand name) instrument.
   G. Scrubber Water Quality: _____% dry solids, particle size; _____% less than 5
      microns, _____% less than 15 microns, _____% less than 25 microns in
      microscopic size and converted to _____% less than 5 microns, _____% less
      than 15 microns, and _____% less than 25 microns on weight basis _____%
      filterable solids, _____ pH.
   H. Cognizant Regulation: Particulate emission parameter must be removed to
      _____ and _____ (gr/scf and lb/hr) for a _____% removal efficiency according
      to _____ (cognizant regulatory authority regulation). Factor-of-
      safety of _____ desired in requirements control equipment manufacturer
      must guarantee. _____ testing procedure(s) must be followed
      for acceptance test and _____ testing procedure(s) must be followed for
      compliance tests. The front half (i.e., nozzle, probe and filter catches) will be con-
      sidered particulate and the condenser catch, consisting of _____,
      _____ & _____ (extraction for hydrocarbon, water
      evaporation after extraction and acetone rinse of condenser) _____ (will
      or will not) be considered as particulate.

(continued)

TABLE 5.1. (continued).

III. **Gaseous Emission Parameters:**
  A. Gaseous Parameter & Concentration: _____ & _____ (ppm on a volume basis = ppmv).
  B. Mass Rate: _____ (lb/hr).
  C. Cognizant Regulation: Gaseous emission parameter must be removed to _____ppmv and _____ lb/hr for a _____ & _____% removal efficiency according to _____ (cognizant regulatory authority regulation).
  D. Evaluation Technique(s): _____
  (process information, similitude, simulation and/or testing by method _____

  _____
  (description of test procedure(s)).
  E. Absorbability: _____ (highly or mildly) absorbable in _____% solution of _____ (reagent) with reaction rate = _____, and _____ lb/hour of reagent required with _____ lb/hour of sludge at _____% solids and _____% filterable solids will result.
  F. Adsorbability: _____ (highly or mildly) adsorbable in _____ (adsorbate). The life of the adsorbate is estimated to be _____ (hours, days, weeks).
  G. Combustibility: _____ (combustible or incombustible) and most suitable for _____ (direct flame, indirect flame, or catalytic) combustion technique. Catalyst poisons and fouling agents are _____ (present, not present) including _____, _____, _____, _____, _____ (list five most severe catalyst contaminants).

IV. **Emission Parameter Handling:** _____ (ton/year) collected at interval of _____ (weight)/_____ (time) and handled in a _____ (slurry, dry, etc.) state. Value of disposable is _____($)/_____ (weight) utilized as _____ (end user) which will account for _____(%) of material produced from the emission parameter collector.

V. **Power Availability:**
  A. _____ (amps) at _____ (volts) _____ (phase) _____ (kHz) available.
  B. _____ (amps) at _____ (volts) _____ (phase) _____ (kHz) required by selected control system.

VI. **Space Availability:** _____ square feet at _____ (ground or elevated) with _____ (process element located below available space for installation of control equipment).

VII. **Financing Availability:** _____ (company, bond sales, commercial bank, etc.) funds at an effective interest rate of _____(%) to be utilized.

VIII. **Esthetic Considerations:** (Steam plumes, black box appearances, neighborhood compatibility, odor, etc.)

68

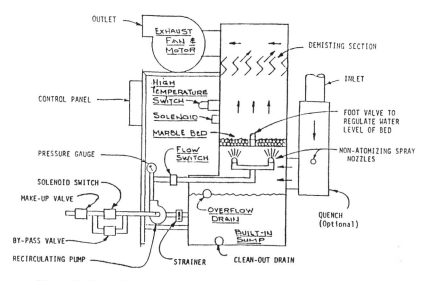

Figure 5.2 Schematic of typical small incinerator scrubber (Vari-systems design).

5.3 COMPARISON OF CONTROL DEVICES

Some of the equipment flow diagrams or sketches for typical wet air pollution control systems are shown in the following figures. Figure 5.2 is a packed bed (marble bed) absorber for gas control and some particulate control. The unique high efficiency catenary scrubber and the collision scrubber are shown in Chapter 7 (Figures 7.8 and 7.7, respectively). These are good on both gases and particulates. Figure 5.3 is a combination baghouse (for particles) and countercurrent spray times (for gases and vapors). Figure 5.4 shows a Venturi scrubber. This may also be used as a quencher to improve control efficiencies of other devices.

5.4 SPECIFICATIONS FOR WET SCRUBBER SYSTEMS

5.4.1 GENERAL CONSIDERATIONS FOR CONTROL DEVICES

A general listing of pertinent topics to construct and building air pollution control devices is in Table 5.2. This extensive listing includes bid evaluation ideas, warranty considerations and the useful data.

5.4.2 SPECIFIC SCRUBBER SPECIFICATIONS

A typical set of specifications for a Venturi/packed tower control system

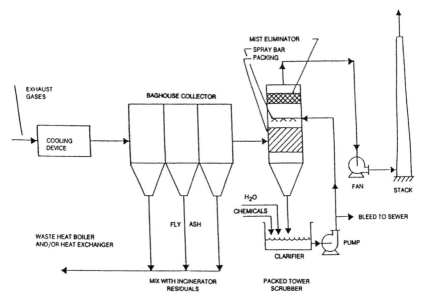

Figure 5.3 Baghouse packed tower air pollution control system.

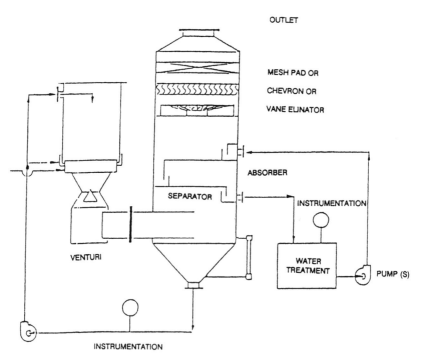

Figure 5.4 Quencher Venturi scrubber.

TABLE 5.2. **Pertinent Topics Relative to Contracting
Air Pollution Control Devices.**

Process Information

Conditions which impact application, size and service *should include:*

- source description
- gas volume
- temperature
- pressure
- particulate loading (gr/scf)
- fuel details
- gas concentration (ppm)
- dust analysis
- particle size
- moisture content

Design Parameters

Specified limitations on Critical Design Parameters

Precipitators

- maximum gas velocity
- minimum specific area
- minimum aspect ratio
- minimum number of fields and bus sections
- maximum migration velocity

Fabric Filters

- air to cloth ratio
- type of cleaning
- bag size
- fabric material
- maximum pressure drop

Scrubbers

- maximum pressure drop type (Venturi, packed tower, etc.)
- scrubbing liquid

Hardware Details

- materials of construction
- sizes, shapes, arrangements
- capacities, code classifications, types

General Specifications

- painting
- thermal insulation
- structural
- electrical
- access
- safety requirements
- state and local codes

Terms and Conditions

- terms of payment
- delivery schedule
- force majeure clause
- supplier scope of liability

(continued)

TABLE 5.2. (continued).

Specifications
- complete
- detailed
- clear
- unbiased
- factual
- no extraneousness
- well-composed
- orderly
- reasonable

Bid Evaluation
- honest
- unprejudiced
- thorough
- technical
- reasonable
- clear
- orderly
- logical

Scope of Supply
Specific, detailed listing of vendor and purchaser supply

Example

Vendor
- double chamber precipitator
- inlet and outlet transistors with gas distribution devices
- access facilities
- support structure
- inlet slide gates
- outlet louver dampers
- inlet and outlet duct work
- expansion joints
- thermal insulation and sliding
- erection

Purchaser
- foundations
- site preparation
- access to two sides of site
- storage area
- dust removal system
- ID fan and stall
- low voltage wiring and conduit
- electrical control room

Specifications
- detailed and exact statement of particulars
- prescribes materials, dimensions and workmanship
- establishes commercial terms and conditions of sale

Guarantees and Warranties
- particulate emission level
- gaseous emission levels
- opacity, carbon monoxide, etc.
- operating characteristics during performance test
- gas distribution guarantee
- extraordinary equipment warranties
- power, water, pressure drop

Bid Evaluation
- preliminary review
- detailed tabulation
- physical arrangement
- final selection
- final negotiation

for particulates/acid gases control is given here under the following twelve topic headings:

(1) General

    a. The air pollution control system should be specifically designed to operate with the incineration and waste heat recovery equipment proposed.

    b. The air pollution control system should be designed to control particulate and acid gas emissions characteristic of the medical waste incineration process.

    c. The air pollution control system should be a wet scrubber design, to include both Venturi throat and packed bed sections. Any systems, wet or dry, that differ from this technology must be submitted by the CONTRACTOR as an alternate bid.

(2) Operating Conditions

    a. The air pollution control system should be designed to accommodate exhaust gases exiting the waste heat recovery system.

    b. The air pollution control system should be designed to operate with an outlet particulate concentration of 0.015 gr/dscf corrected to 7% O_2. The inlet particulate concentration should be assumed as 0.20 gr/dscf corrected to 7% O_2 as required in Section II of these specifications.

    c. The air pollution control system should be designed to operate with a hydrogen chloride emissions rate of 2.2 lb/hr.

    d. The air pollution control system should be designed to operate during periods when soot blowers included in the waste heat recovery system are activated.

    e. The air pollution control system should be designed to operate with no visible stack plume. The CONTRACTOR shall submit, along with his bid, documentation of the method used to achieve this requirement.

(3) Alternate Bid

    a. As an alternate bid, the air pollution control system shall be designed to operate with no liquid discharge. The CONTRACTOR shall submit, along with his bid, documentation of the method used to achieve this requirement.

(4) Quench

    a. A quench section shall be included for the purpose of cooling exhaust gases exiting the waste heat recovery system.

    b. The quench section shall be constructed of a high nickel alloy, such as Inconel 625 or Hastelloy C276, to be resistant to gas contaminant corrosion at elevated temperatures.

(5) Condenser

   a. A section of the system shall be included to provide for removal of moisture contained in the exhaust gas. This condenser should be used to minimize the visible vapor plume exiting the exhaust stack.

   b. The condensing section shall be constructed of a high nickel alloy, such as Inconel 625 or Hastelloy C276, to be resistant to gas contaminant corrosion at elevated temperatures.

(6) Venturi Throat Scrubber

   a. A Venturi throat section shall be included to provide control of particulate matter contained in the exhaust gas.

   b. The Venturi section shall be equipped with an automatically variable throat, so as to maintain a constant incinerator primary chamber draft at varying flow rates. Minimum pressure differential capability shall be 30 in. w.g.

   c. The Venturi section shall be constructed of a high nickel alloy, such as Inconel 625 or Hastelloy C276. These materials are resistant to gas stream contaminant corrosion at elevated temperatures and concentrations.

(7) Clarifier

   a. A clarifier shall be provided to remove particles contained in droplets that are entrained in the gas flow exiting the Venturi section.

   b. The clarifier shall be constructed with a shell of high-grade reinforced fiberglass with chemical resistance to all the gas stream contaminants.

(8) Packed Bed Scrubber

   a. A packed bed section shall be included to provide control of acid gas contaminants (primarily hydrochloric acid) contained in the exhaust gas.

   b. The packed bed section shall be constructed with a shell of high-grade reinforced fiberglass with chemical resistance to all the gas stream contaminants and the reacted products with result from Sodium Hydroxide or Sodium Carbonate neutralization at temperatures up to 200°F.

   c. The packed bed section shall be packed with dump type packing elements designed for the purpose of maximizing the absorption and scrubbing surface area.

   d. The packed bed section shall be equipped with a feed and recirculation loop with flow meter for caustic addition to the system.

Scrubber blowdown effluent and recirculated water shall be maintained at a pH factor of between 6.85 and 7.0.

e. The packed tower shall be equipped with a chevron or equivalent type mist eliminator with fresh backwash capability that can be regulated on a time cycle.

(9) Caustic Addition System

a. A chemical feed system shall be provided for addition of caustic soda to the packed bed section. This shall include a storage tank, feed pump, and associated piping.

b. The storage tank shall be designed to contain sufficient capacity for seven (7) 24-hour days of system operation. The tank shall be of HDPE materials and of vertical cylinder type design.

(10) Control System

a. A single control panel shall be provided in freestanding configuration to contain all of the instrumentation required for the system. Motor starters shall be located in the panel and shall be isolated from the low voltage control circuits as much as possible. The control panel shall include a pH controller, a differential pressure controller, differential pressure indicators across the Venturi section, the packed tower absorption section, and the demister section. Alarms shall be provided on high and low pH, low differential pressure, excessive temperature in the packed tower absorber, low liquid level in the recirculation tank, high liquid level in the recirculation tank, and low liquid flow rate to the Venturi scrubber and packed tower absorber.

b. The control panel should include the following:

1) Enclosed steel freestanding cabinet
2) Baked enamel white inside, gray outside finish NEMA 4 enclosure with a cabinet key lock
3) All controls, instruments, and other equipment shall be flush mounted at factory and assembly tested prior to shipment.
4) All controls, instruments, and other equipment shall be identified by laminated individual nameplates.
5) All components within panel shall be mounted on a special perforated metal subplate, with color coded wiring concealed in plastic troughs, and with numbered terminal strips for identification or external connections.
6) Sensing lines from panel mounted devices shall be piped to a terminal point on panel.
7) Complete wiring diagrams showing panel terminal connections and interconnections with equipment.

(11) Induced Draft Fan

   a. An induced draft fan shall be provided to maintain proper draft throughout the system.

   b. The fan should be designed to compensate for variations in outlet pressure from the Venturi throat section. The CONTRACTOR shall provide, along with his bid, documentation of design features included to accomplish this.

   c. The induced draft fan shall be constructed with the wheel made of the same alloy as the quench and Venturi, and the housing shall be constructed of rubber lined mild steel.

   d. The fan shall be constructed with the fan wheel hub extended through the fan backplate and a Teflon seal installed on the hub.

   e. Fan shall be statically and dynamically balanced.

   f. Fan shall be radial blade centrifugal. Backward inclined, and radial tip designs are not acceptable.

   g. A drain connection shall be provided consisting of a threaded pipe coupling welded to the lowest point on the scroll to allow condensate or other liquids to drain.

   h. An access door securely held by quick release latches shall be supplied for easy access to the wheel and housing interior for inspection or cleaning. The door shall be shaped to conform with the scroll curvature and gasketed to minimize gas leakage.

   i. The fan shall bear the AMCA seal. The rating of the fan shall be based on tests conducted in accordance with AMCA standards 210 and comply with the requirements of the AMCA certified ratings program.

   j. The CONTRACTOR shall submit, along with his bid, a complete set of draft calculations to demonstrate correct sizing of the I.D. fan. All assumptions and results should be clearly stated.

(12) Stack

   a. An exhaust stack of CPVC construction shall be provided for venting gases exiting the air pollution control system to the atmosphere.

   b. The stack shall be 85 feet in height and shall penetrate the roof adjacent to the bypass stack, as shown in Figure 1 of these specifications.

   c. This is the main waste to energy system stack and shall have test ports, platforms, ladders, etc., in accordance with existing requirements.

   d. The maximum inside diameter of the CPVC stack shall be 2 feet.

TABLE 5.3. Comparison of Air Pollution Control Systems for Industrial Boilers.

Parameter	Wet/Dry Scrubber	Dry/Dry Scrubber	Venturi Packed Tower	Baghouse Packed Tower	Collision Scrubber	Wet Electrostatic
Particle capture efficiency, %	99.9+	99.9+	99+	99.9+	99+	Variable, 99–99.9
Particulates out	0.015 gr/dscf	0.015 gr/dscf	0.03* gr/dscf	0.01 gr/dscf	0.01 gr/dscf	0.03–0.01
Acid gas collection eff., %	85–95	75–90	98+	98+	98+	
Plume	None visible	None visible	Visible	Visible	Visible	Visible
Space requirements, feet	40 × 60	40 × 60	20 × 40	30 × 40	30 × 40	30 × 40
Residuals	Wet	Dry	Wet	Dry (baghouse) Wet (packed tower)	Wet	Wet
Typical treatment chemicals	Limestone solution	Dry limestone	NaOH, lime or limestone solution	Liquid NaOH lime or limestone solution	Liquid NaOH or lime solution	Liquid NaOH lime solution
Required ancillary equipment	•Limestone storage •Mixer	•Limestone storage	•Chemical storage •Clarifier •WWT •Slaker	•Chemical storage •Clarifier •WWT •Slaker	•Chemical storage •Clarifier •WWT •Slaker	•Chemical storage •Clarifier •WWT •Slaker

*Venturi Scrubber/packed towers will typically achieve 0.03 gr/dscf at a pressure drop across the Venturi of approximately 30″ water column. Levels of 0.01 gr/scf can be achieved at higher pressures.

5.5 TYPE OF CONTROL SYSTEMS

Systems include the basic devices plus the ancillary equipment. Basic devices are for control of particulates and gases. Examples as noted in the Introduction of this chapter are

(1) *Particulates* (including ash, metals, condensibles, mists, etc.)
  - cyclones (mechanical collectors – used as precleaners for high dust loadings such as stoker coal burning systems)
  - Venturi type scrubbers (e.g., Anderson, Calvert, ACI)
  - baghouses

(2) *Acid gases* (SO_2 and HCl)
  - packed towers
  - dry scrubbers
  - other type absorbers

(3) VOCs and NO_x
  - absorbers
  - wet/dry and dry/dry scrubbers
  - catalytic oxidation/reduction

Most of the air pollution control systems contain other ancillary equipment, including:

  - pre-quench/cooling
  - chemical storage
  - disposal of residue
  - water treatment (for wet scrubbers)

Also, plan on environmental monitoring systems, which will usually be required for opacity, O_2, CO_2 and CO.

Table 5.3 compares several air pollution control systems. It lists typical particle and gas control efficiencies, shows example space requirements and notes ancillary equipment requirements.

CHAPTER 6

Cyclonic Dust Collectors

DAVID L. AMREIN

6.1 INTRODUCTION

CYCLONES are the most economical and reliable means for the removal of particulates from a gas stream. They also hold the dubious distinction of suffering from the most misunderstandings and subsequently, poor engineering of any type of industrial dust collection equipment.

Before discussing misunderstandings and myths surrounding cyclones, it is important to first define what a *cyclone* is. In the most general sense, a cyclone is an inertial separation device that utilizes a circular primary flow pattern. A cyclone functionally produces a high velocity vortex of particle laden gases. Centrifugal force hurls the dust particles towards the cylindrical walls of the collector. Figures 6.1–6.5 conceptually depict different types of "cyclones." Many other devices that may be defined as cyclones also exist but are not shown. For our discussion, the device shown in Figure 6.1 shall be considered when we use the word *cyclone*. The most accurate description of this device would be *"a reverse flow cyclone with involute inlet."* It is called "reverse flow" because the gas enters involutely through the top of the collector, spinning and traveling downward. The gases then reverse direction 180 degrees, spinning in the same direction. "Involute inlet" describes the manner of gas entry into the cyclone, which is the cause of the spinning motion. The reverse flow cyclone is generally considered the most efficient configuration for dry cyclones, since the particles are conveyed to the collection point (cyclone discharge) by the gas flow.

The general primary flow pattern of a cyclone is depicted in Figure 6.6 for concept purposes. The position, magnitude and the direction of the conceptual flow lines may vary considerably from one cyclone design to

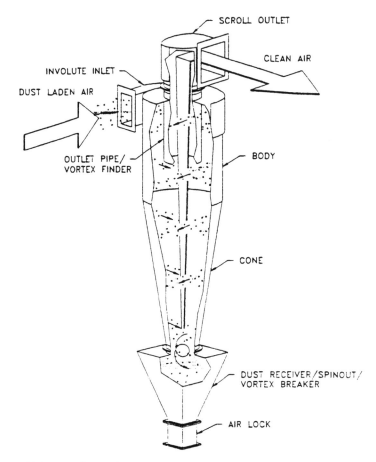

Figure 6.1 Reverse flow cyclone. (Note: The cyclone dust collector consists of a cylindrical/conical shape with no internal moving parts. Dust particles are separated from the air stream by centrifugal force created by the air's tangential velocity.)

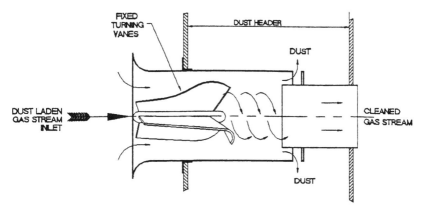

Figure 6.2 Fixed impeller straight-through cyclone.

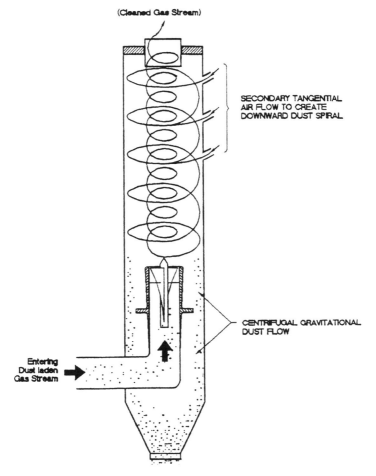

Figure 6.3 Straight-through cyclone with reverse dust flow.

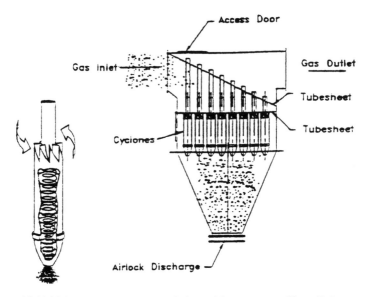

Figure 6.4 Multiple conventional cyclone design axial entry type with preliminary settling chamber.

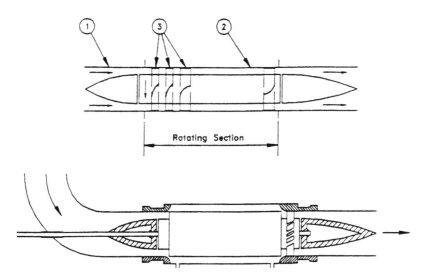

Figure 6.5 Turbo-compressor air cleaner. Item 1 is a fixed cylindrical inlet duct, item 2 is the rotating inner sleeve of the centrifuge, and item 3 is compressor turbine blades. Clearances between the outer casing of the centrifuge and the fixed inlet and outlet ducts permit free rotation of the centrifuge and the discharge of the liquid film.

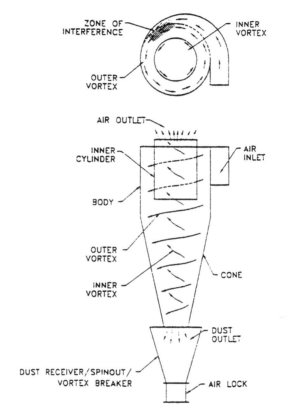

Figure 6.6 Flow pattern in a conventional, reverse-flow cyclone.

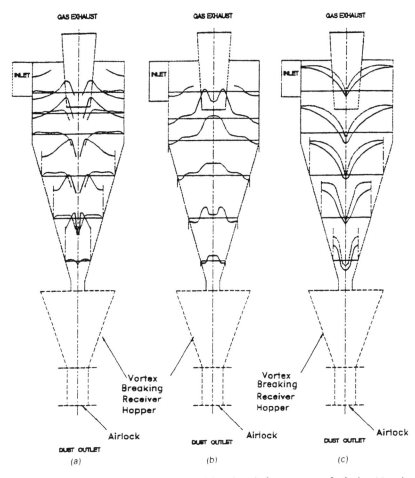

Figure 6.7 Relative variation of tangential, radial, and vertical components of velocity: (a) variation of tangential velocity and radial velocity, (b) variation of vertical velocity, and (c) total and static pressures.

another. In general, the primary flow pattern within a cyclone is a double vortex. The gas entering the top spins down along the outside walls of the cyclone, then reverses in the same direction, upward along the axis of the cyclone. Figure 6.7 shows a typical representation of the secondary flow patterns (vertical and horizontal components without spinning) within cyclones. These representations are "telling" in one regard: in that they show the flow pattern to be non-absolute with much looping and recirculation of gas flow. Although this picture is less clear, it is more realistic than the concept of definite and distinct flow patterns often represented in cyclones.

6.2 THE INLET

The inlet design can have a dramatic effect on the final results of the collector, because it controls the velocity and disposition of the particulate coming into the unit. The inlet is involute in configuration as it directs the gas stream, laden with particulate, around a central axis formed by the outlet pipe/vortex finder.

Differences have been noted by various cyclone manufacturers about inlet configuration. It has been noted that the inlets can be configured as rectangular or round, and located as involute, axial or tangential (see Figure 6.8). Most experts agree that the involute design can offer more overall advantages than the tangential design.

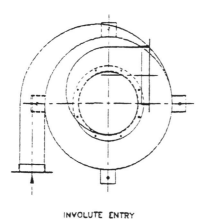

INVOLUTE ENTRY

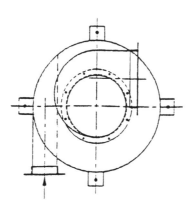

TANGENTIAL ENTRY

Figure 6.8 Primary cyclone inlet configuration.

Involute Advantages	Tangential Advantages
1. Smoother transition into the body offering a minimum of inlet resistance 2. Less abrasive transition 3. Less likelihood of abrading the outlet pipe 4. Less turbulence 5. Less pressure drop for a given flow 6. Better collection efficiency for a given energy consumption (pressure drop)	1. Less materials of construction 2. Ease of construction 3. Better collection efficiency for a given flow rate

6.3 THE BODY SECTION

Assuming a proper inlet condition has been met, the particulate laden gas enters the upper body section where centrifugal and gravitation forces begin to take effect. It should be noted that centrifugal forces, being much greater than gravitational forces, allow collection of a much finer particulate than a settling chamber or gravity dropout box. The importance of the diameter and length of the body, and its relationship to the inlet and outlet, cannot be overstated. The size and length of the body, relative to the unit velocity and outlet pipe configuration, establish the vortex strength and centrifugal forces within the spinning gases. As the gases spin around the wall and progress axially downward, the particulate presses centrifugally outward while the gases begin to seek the central vortex, which will exit by way of the outlet pipe. Figure 6.7 conceptually depicts the variations in velocity components and pressures as expected in an involute reverse flow cyclone.

Primary and secondary gas flow patterns are prevalent also and are indicated conceptually on Figures 6.6 and 6.9.

6.4 THE CONE SECTION

The conical part of the cyclone plays a major part in

(1) Confining the vortex to the center of the cyclone
(2) Reducing the effects of the spinning gases (see Figure 6.10)

The cone is an important transitional area, where the efficiency of a cyclone can be dramatically altered by small unfavorable circumstances

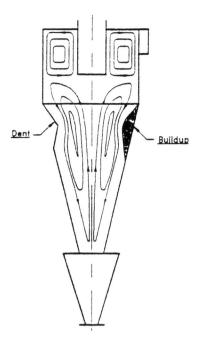

Figure 6.9 Secondary flow patterns caused by dents and buildups.

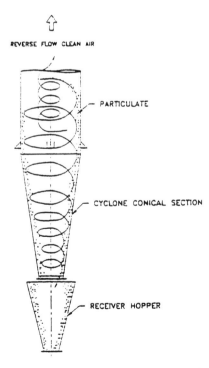

Figure 6.10 Cyclone conical section configuration.

that can easily be overlooked by the equipment operator. Reentrainment in this area of the vortex can occur for a variety of reasons:

(1) Bridged material impeding proper discharge
(2) Secondary air leakage through the discharge
(3) Unfavorable eddy currents due to dents in the side walls or material buildup
(4) Possible effects due to the lack of a proper electrical ground
(5) A short body or cone section
(6) An undersized discharge

The cone's major function is to control the delivery of material to a central point of discharge. Its length can play a very important role relative to retention time, pressure drop, and vortex strength at the point of discharge.

6.5 THE RECEIVER OR SPINOUT

The receiver or spinout is a transitional airtight area at the bottom of the cone that any proper cyclone design includes (see Figure 6.11). If designed

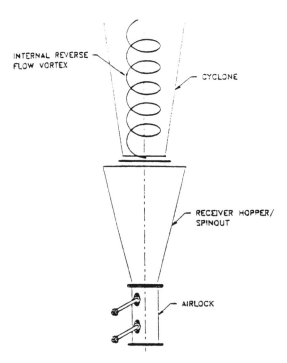

Figure 6.11 Receiver/spinout vortex breaker.

correctly, conditions at the bottom of the receiver (prior to the airlock) are much improved over the entry conditions from the cyclone discharge. This part of the equipment allows the particulate to spin out and avoid reentrainment into the upward spiraling inner vortex.

6.6 THE OUTLET PIPE/VORTEX FINDER

The outlet pipe design ranks nearly as importantly as the inlet in importance toward design consideration. An outlet pipe with a hole worn in it will create a major deterioration in total collection efficiency of the equipment.

A proper outlet pipe extends through the roof in the central axis of the equipment, down into the central body, to a point below the bottom of the inlet. This part of the cyclone helps eliminate reentrainment influences, as particulate initially enters the cyclone collector, by causing the gas flow to travel downward. Many cyclones include manufactured cylindrical outlet pipes that are sized for a controlled velocity and static pressure drop. Manufacturers of ultra high efficiency designs can also utilize a very specific tapered outlet pipe to assist with energy regain, thus aiding the ability to provide higher efficiencies at lower pressure drops. The physical size of the outlet pipe that controls the outlet velocity is as important to maximum efficiency design as the inlet configuration.

6.7 CYCLONE MYTHS

Many myths are associated with cyclone performance, with the most prevalent being that cyclones are rated at the bottom of the list of dust collection efficiency. In many applications high efficiency cylones are acceptable replacements for more exotic collection equipment for emission control and product recovery.

A common myth encountered in day-to-day discussions is the assumption that a specific number of revolutions (or turns) is required to achieve a specific efficiency. The turn theory is based on conclusions that visible bands seen on the interiors of some cyclones indicate that particulate has entered the equipment, followed the visible path and exited the discharge. It is a fact that much of the particulate and gas travels a much different path, requiring many times the turns indicated by the average visible spiral.

One of the great myths relative to cyclones is the one that insinuates that a receiver hopper or airlock is not necessary if the cyclone utilizes a blower prior to the cyclone. Whether the internal "tornado" is created by positive or negative in-draft, it is still a tornado (see Figure 6.12). This

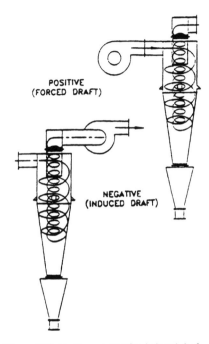

Figure 6.12 Positive or negative induced draft.

means it can still have the tendency to act as a vacuum cleaner and reentrain material at the discharge, much like a real tornado that lifts cars, houses, and the sod from the ground. This is not to say that influences cannot be designed in, to create a positive pressure at the discharge, but that a cyclone may not provide satisfactory results in a positive system without a proper receiver hopper and airlocked discharge unless great care is taken in the design.

Thousands of cyclones are being operated throughout industry everyday that provide much less than optimum results because the operator thinks that one cyclone is the same as another. Many cyclones operating at less than 90% total collection efficiency could be upgraded with a cyclone capable of efficiency exceeding 99%.

6.8 COST CONSIDERATIONS

When comparing a cyclone dust collector to a wet scrubber, electrostatic precipitator or a fabric filter, the end user should take in consideration the initial capital expenditure, installation costs, operating costs, and maintenance costs relative to the collection results achieved (see Table 6.1).

TABLE 6.1. Comparisons of Collectors (for specified PSD and DATA).

	Percent Removal Efficiencies at 6″ ΔP Overall for Particles			Approx. Percent Total Collection Efficiency	Initial Relative Cost	Relative Operating Cost	Relative Maintenance Cost
	9 μm	4 μm	2 μm				
Cyclones							
Ultra high efficiency	99.3	94.5	61.0	96.4	Low/Med	Very Low	Very Low
Standard efficiency	80.5	51.5	25.4	83.9	Low	Very Low	Very Low
Scrubbers							
Venturi w/mist eliminator							
6″ ΔP	99.8	91	36.7	94.9	Low/Med	Med/High	Med
22″ ΔP	99.9	99.8	99	99.2	Low/Med	High	Med
Electrostatic Precipitator	99.99	99.9+	99.9	99.9+	Very High	Very Low	High
Baghouse	99.99	99.9+	99.9	99.9+	High	Low	Moderate

6.9 SPACE REQUIREMENTS

Standard efficiency general duty cyclones are used throughout most powder and bulk industries. These cyclones tend to be large diameter units but short in stature. These cyclones tend to be adequate for collection of most $+325$ mesh ($+44$ μm) sized particulate. When high and ultra high efficiencies are required on finer particulate, especially those finer than 450 mesh, single cyclones tend to be much longer and require considerable height. One of the ways to reduce height, when not available, is to cluster cyclones as dual, quad, etc., arrangements in parallel (see Figure 6.13). Multiple arrangements, although more costly, can provide higher efficiencies at the same pressure drop of an equivalent single arrangement.

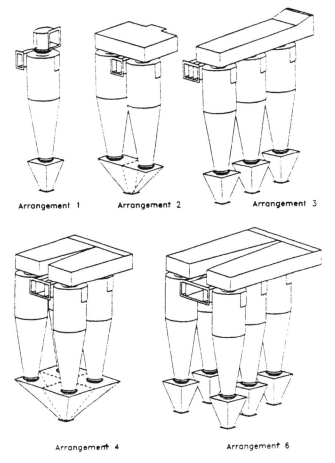

Arrangement 1   Arrangement 2   Arrangement 3

Arrangement 4   Arrangement 6

Figure 6.13 Single to multiple cyclone arrangements.

6.10 CYCLONES COMPARED TO OTHER CONTROL DEVICES

In the comparison of cyclones, scrubbers, electrostatic precipitators, and baghouses, it must be understood what is being compared. Referring to Table 6.1 one can note the relative comparisons in efficiency and costs. Cyclones, being aerodynamic collectors with no moving parts provide the easiest consideration relative to maintenance. If the user opts for materials of construction, that will minimize the potential for accelerated abrasion, when abrasive materials are to be collected, then the user should be a happy user.

Other collection devices must deal with more detailed considerations. For example, scrubbers must deal with the water system design and the disposition of the wet collected materials. Electrostatic precipitator considerations must deal with maintenance and costs to maintain rapper systems, electrical transformers, and costly material recovery systems that are required on the multiple field hopper systems. Baghouse considerations include the effects of shutdown for bag filter repair, replacement, or cleaning.

6.11 CYCLONE TROUBLESHOOTING

It must be understood that a cyclone's performance is based on the specific inlet conditions provided by the customer initially. Any changes to the input for the incoming gases and/or particulate will provide a change in the discharge and outlet results. Therefore, when troubleshooting an existing cyclone that appears to be operating out of specification, it is important to determine if the operating conditions are as specified.

The primary fault detectors are the pitot tube and the manometer or U tube. The pitot tube is utilized to ascertain the velocity/volume data in the inlet and outlet ducts. The manometer/U tube is utilized in conjunction with the pitot tube, and also to ascertain the inlet static pressure and the outlet static pressure to determine if problems exist before, after or within the cyclone.

It is extremely important that the entering gas flow volume equals the exiting gas flow volume, to establish that no leaks exist. Leaks can cause severe negative influences on cyclone performance. If this does not provide enough indication of a fault, then more sophisticated means are required to obtain an isokinetic test.

Particulate materials that tend to be sticky, creating potential bridgeover conditions can alter the optimum flow patterns within the equipment, causing more severe problems. The solution utilized at most industrial plants to free a bridgeover can ultimately create more problems. How many times has an operator or maintenance man pounded on the light gauge construction with a mallet or hammer and caused dents in the equipment? A dent can easily redirect the particulate laden gas stream to the inner vortex, thus reducing efficiency.

All particulate, as it passes through a cyclone interacts with itself during its journey. The higher the loading, the more interaction occurs creating the potential for a static electric discharge. Entire articles have been written on this subject. It is noteworthy to point out that, in many cases, static electricity causes some materials to collect better when the equipment is grounded, while others collect better when the equipment is not grounded. Some particulate takes the static charge better than others.

Some bridging conditions may not be caused by static electricity but by wetting conditions caused by condensation. If a cyclone has been utilized in a hot gas system and a proper insulation has not been installed, then it is not uncommon to create a condensation problem where water literally pours from the discharge. A cyclone retains a gas molecule much longer than would normally be expected, due to the rotation and retention time.

An optimum cyclone design is contingent upon accurate input. Changes in input will ultimately affect the output of the cyclone.

The data input necessary for a proper design includes accurate gas and particulate information.

6.12 GAS CONDITIONS AT COLLECTOR INLET

_____ PSIG _____ °F or Volume _____ ACFM, Pressure _____ PSIA, Temp. _____ °C

Moisture content (Specify by weight or by volume.)

Attach gas analysis if other than air (Specify by weight or by volume.)

Particulate description _____

Specific gravity _____ Bulk density _____ lbs/cuft

Is material corrosive? _____ abrasive? _____ sticky? _____ explosive? _____ toxic? _____

Dust load at inlet side of collector _____ lbs/hr or grain/ACF

Attach particle size distribution _____ Test method _____

With this knowledge established, the remaining factors that enter into the selection process are

Maximum allowable pressure drop _____

Operating pressure _____

Desired collection efficiency _____

Size restrictions _____

A few manufacturers have the capability of performing a computer analysis on any existing cyclone, based on accurate data input from the user.

This analysis can be used to establish the degree of improvement that is possible, utilizing a modern high efficiency model.

Cyclones utilized as secondary and primary collectors are desirable as the least expensive dry collector, considering capital cost, operational and maintenance cost.

Cyclones are generally utilized more as industrial process collectors than as pollution control devices. Many states refuse to grant a permit for a cyclone to industry for pollution control unless it is used as a precleaner to a primary fabric filter collector.

Although cyclones can be designed to provide invisible emissions in many industrial applications, few manufacturers have developed the technology to the degree required.

CHAPTER 7

Principles of Wet Scrubbers

HOWARD E. HESKETH

7.1 INTRODUCTION

THIS chapter deals with wet scrubbers and related facilities. We will be discussing principles of atomization, collection mechanisms and what to do with hot and cold gases. In general, types of scrubbers, particulate collection with and without gas absorption and mist elimination are included. We'll proceed to discuss operation, giving recommended procedures that have been developed by this author and others that will help you, the reader, predict characteristics of operation for specific devices. The Venturi scrubber has been developed most fully, and will be discussed relative to pressure drop, throat length, and finally correlating these with particle removal efficiency. Gas removal efficiency is included. Material of construction as it relates to the operation of these devices is also presented.

In contrast to cyclones that are for particulates only, scrubbers do a good job for both particles and gases. This assumes that they are properly designed, operated and maintained. Baghouses and precipitators are normally for particulates only, with the exception that you can combine these with wet/dry spray/dry operation or dry/dry injection.

7.2 SYSTEMS

Before discussing devices, it is important to look at scrubbing systems. One of the first parts should be the quencher. If you're quenching for a wet scrubber, the quench attempts to achieve full adiabatic saturation of all material present. If you're conditioning for a dry scrubber absorber, the type of quench differs in that you only condition a gas going to a dry ab-

TABLE 7.1. **System and Component Notes.**

Flue Gas Cleaning
  Cyclones—particle control only
  Wet scrubbers
    Spray absorber towers—gases
    Venturi scrubber—particles
    Collision scrubber—particles
    Catenary scrubber—both
  Baghouses—"particles" normally
  ESPs—"particles" normally
  Dry scrubbers—both
    wet/dry spray dry absorber (SDA)
    dry/dry dry injection
  Combinations of systems
1st step is usually to quench hot gases to <200ºF.
Acid gas control in absorbers requires 5–10 sec residence time.
Particle emissions are <1% smaller than 0.35 μm and 90% less than 25 μm with mass
  emissions allowable usually ~0.08 grains/dscf.
The system produces ash, fly ash, and wet sludge.

---

sorber. You want to ultimately reach a temperature 20–30 degrees *above* the dew point. You want to approach the dew point, but you do not want to reach it. Table 7.1 shows application of some systems and some notes on how long it may take to achieve quenching jobs. The times noted are maximum times required. It could be possible that a couple of seconds might be adequate to do the job. The longer the contacting time between the gas and the quenching liquid, the better the results will be.

7.3 PARTICLE SIZE

It is critical that we know particle size distribution of the material to be scrubbed. These data can be conveniently plotted on log-probability paper as shown in Figure 7.1. This figure shows a typical particle size distribution for particles from incinerator systems. The inert material from the incinerator has a mass mean diameter of approximately 13 microns. For comparison, condensation aerosols from the incinerator (condensing toxic metal emissions, in particular) may have a mass mean diameter of 0.46 microns. This shows that, as typical of many systems, there is more than one type of dust in the mixture. Here is evidence of direct emissions plus condensation emissions. Condensation emissions are frequently much smaller in size, so they are much more difficult to collect.

7.4 COAL-FIRED BOILER

Table 7.2 is an example of operating parameters for a 25-MW coal-fired

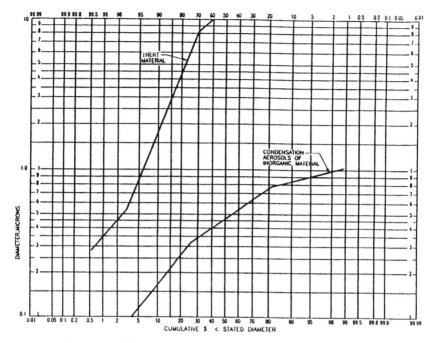

Figure 7.1 Typical size distributions of incinerator particle emissions.

boiler. Specifically, these are based on bituminous, pulverized coal with a good heating value, 10% ash, high sulfur, an annual load factor of 80% (which is slightly higher than average but, nevertheless, typical of many systems), boiler firing rate, and 85% removal of the SO_2. Twenty-five percent of the ash goes to the bottom, and 75% goes out as fly ash. Sulfur re-

TABLE 7.2. **Example of Coal-Fired Boiler Operating Parameters, Including Emission Control Systems.**

Boiler capacity—25 MW
Fuel—pulverized coal, bituminous
Heat valve—10,400 Btu/lb
Ash content—10%
Sulfur content—2.5%
Annual load factor—80%
Boiler firing rate—9500 Btu/Kwhr
Required sulfur removal 85% scrubber (double alkali)
Ash distribution—25% bottom, 75% fly
Sulfur retained in ash—10%
Bottom ash bulk density—50 lb/ft³
Fly ash bulk density—75 lb/ft³
FGD sludge bulk density—95 lb/ft³ (at 60% solids)

TABLE 7.3. **Weights and Volumes of Solid Waste (25-MW Coal-Fired Boiler).**

Waste	Weight (Tons/yr)	Volume (Acre-ft/yr)	(Acre-ft/30-yr)
Bottom ash	1,980	1.82	54.64
Fly ash	5,945	3.64	109.14
Scrubber sludge (60% solids) Utilizing limestone	23,545	11.38	341.40
Reagent consumption	4,300 tons/yr limestone		

tained in the ash is shown as 10%. This is high, but it could indicate a very alkaline ash (5% sulfur retention would be more typical). Bottom ash bulk density, fly ash bulk density, flue gas, and sludge bulk density values are shown. Each of these would be products from the wet scrubbing system if, indeed, we did use wet scrubbing to achieve the control indicated.

When you use a wet scrubbing system, you accumulate waste at an increased rate. A typical system is shown in Table 7.3. Note there are three times more tons per year of fly ash than bottom ash and another fourfold increase in mass due to production of scrubber sludge. The big question becomes: Where do you put all of this "stuff"? If you put it on land, it could cover many acres, several feet deep each year for only a 25-MW system. Over thirty years of operation, a tremendous amount of waste could be produced. In these waste figures are included unreacted limestone (unreacted lime) and free water (60% solids means the balance is 40% water). This waste issue is an important factor to consider with wet scrubbing systems.

7.5 SCRUBBER CHARACTERISTICS

On the other hand, there are advantages to a wet scrubbing system. It is good for particles and gases, including hot gases, odors and sticky material. Filter bags will burn out if you have too hot a temperature or will plug with sticky material. With an ESP, you need to be concerned with dust resistivity and plugging. Dust resistivity in an ESP could be undesirable if you didn't have the proper composition, temperature and conditioning. Scrubbers are usually very good for sticky type material.

Scrubbers may be expensive to operate, because they could require a high pressure drop, similar to a fabric filter. High energy scrubbers are those that work at a relatively high pressure drop. High pressure drop is synonymous with the high operating cost. On the other hand, the pressure drop of a dry scrubber with a fabric filter system could be as large or larger than the pressure drop of a wet scrubber. The wet scrubber has the advantage of cooling unsaturated gases and providing some of the needed time to reach condensation conditions. However, this advantage results in a wet product and the whole control system is more subject to corrosion and erosion. These systems deal with acids, bases and fly ash, which are very rough materials to handle.

Scrubbers can be plugged. In the early days of flue gas desulfurization, one of the problems was that the scrubbers plugged. Indeed, they do if they are not designed and operated properly.

What you are doing, in many cases, is taking a gas, and reacting it with a chemical to produce a precipitate (solid particles). This means that the liquid phase is a solution with solids present. When you saturate (or supersaturate) the solution, some of it will precipitate out, and precipitate could deposit at the wrong place. In a scrubbing system, you must provide time for precipitation to occur, and you want to control the place that it occurs, so that it does not deposit on the internals of the scrubber.

As noted, energy requirements could be high. Also, if nitrogen was not present, the gas pumping costs would be much lower. Here is a relationship between fan, horsepower, pressure drop, and volumetric flow rate at 60% fan efficiency:

$$\text{Blower HP} \cong (3 \times 10^{-4})(\Delta P)(Q)$$

where ΔP = inches w.g. and Q = acfm.

Fan energy required is typically ~ 5 hp/1000 acfm. As a rule of thumb, large flue gas desulfurization systems require 1.5–5% of the system energy output just to keep them operating. The newer novel type of scrubbing systems have lower pressure drops, and are more efficient, but they are more expensive to purchase. In other words, the capital cost could be greater, but they may give a much lower operating cost.

7.6 ATOMIZATION

You cannot have good wet scrubber operation without good atomization of the scrubbing liquid. Atomization provides the targets for the particle to collect upon. These targets are basically about 50–75 microns in size. They are quite large in comparison to the dust, the dust typically having, as indicated, a much smaller mass mean diameter. We have different ways of atomizing this liquid: it can be pneumatic or mechanical. The liquid not only provides a droplet target, but if absorption is required, a lot of surface area must be available. So targets with a lot of surface area are required to

TABLE 7.4. Spray Nozzle Examples.

Nozzle	Orifice φ, inch	Type	Capacity in GPM @ psi					Spray Angle in Degrees @ psi			
			3	7	20	60	80	7	20	60	80
1/8 ks	0.073	Flood jet, flat spray	0.27	0.42	0.71	1.2	/	114	128	142	/
3/8GG9.5	7/64	Full cone	0.54	0.80	1.3	2.2	2.5	45	50	/	46
062FH5	0.062	Flat spray, high velocity, replaceable orifice	/	/	0.38	0.66	0.76	/	70	84	/
ST48XP	3/4	Full cone, large free passage, abrasion resistant	/	24	67	117	135	either 90° or 120°			

Vendors Nos. are one of: Spraying Systems Co., Wheaton, IL; Spray Engrg. Co. (Spray Co.), Burlington, MA; Wm. Steiner Mfg. Co., Parsippany, NJ; and Bete Fog Nozzle, Inc., Greenfield, MA.

give good absorption. Absorption will be directly related with the surface area of the liquid introduced. Spray nozzles can be used to do this atomization, including mechanical or two-phase (gas plus liquid). Table 7.4 shows examples of different spray nozzles.

Spray nozzles are critical components in the system and must be protected. We noted the fact that usually wet scrubbers contain abrasive, erosive materials. Care must be taken to prevent corrosion, erosion or other destruction of the nozzles. Teflon is a good material of construction for some nozzles applications. Teflon is a plastic type material and, seemingly, would wear quickly. However, in some instances, it does not wear quickly because it has a very low coefficient of friction. Other typical construction materials for nozzles are stainless steel, ceramic, rubber-lined, Hastelloy, inconal and others.

Upstream from any pump and before any nozzle system, there should always be a strainer. The strainer is in the line to keep big material from plugging up the nozzles. You should use blow-through, flow-through strainers. These can be blown down during operation and can always be kept clean. However, always plan for the inevitable. A worker can drop his wrench, and it would end up in the pump if you don't have the strainer. Nuts, bolts, hard-hats, anything. As an extreme example, a certain utility is located adjacent to a nice big golf course. The seventh hole is not too far from the scrubbing system thickener. The golfers often try for a hole-in-one in the thickener. Imagine what a golfball does to a positive displacement pump. So, don't plan on the uneventful; anything that can take place, will.

Spray nozzle data such as given in Table 7.4 can be used to calculate scrubbing liquid rates. If you have flow data for the spray nozzle, and you know the nozzle pressure and number of nozzles, you can estimate the liquid quantity being sprayed into the system. It is imperative that you have enough spray to do the job, and on the other hand, it is expensive to put in more than you have to.

7.7 PARTICLE CONTROL MECHANISMS

In particle control using a wet scrubber, inertial impaction is the dominant procedure. Interception is a bonus, and diffusion that exists, is very weak. Phoretic forces are also weak forces, but they can be highly significant. Gas absorption requires that the pollutant gas move through the carrier gas to the surface of the liquid, pass through two films (gas film, liquid film) and move into the liquid surface, then move from the surface into the medium of the liquid that is doing the collecting. Chemical absorption may also occur. This means that a chemical reaction occurs to remove the absorbed pollutant by converting it into an entirely new chemical species.

Quenching, prescrubbing, and flux-force condensation are all essentially synonymous, relative to a wet scrubbing system. These will be discussed later in this chapter.

Figure 7.2 is the familiar diagram of a target and particles to be collected. In our case the target is a little droplet of liquid that has been atomized. Gas approaches the target from the stagnation line and either goes above or below the target. If the particle is capable of traveling with the gas, it will escape capture on the target. If a particle is close to the stagnation line or big, it will not be able to elude capture. If the velocity difference is large enough, the particle will impact on the target. That is inertial impaction. The impaction parameter, K_I, is shown in Figure 7.2. Magnitude of an impaction parameter is related directly to the square of the diameter of the particle being collected, directly with the particle velocity difference (velocity difference between the particle and the target) and indirectly with the target size. Interception (also shown in Figure 7.2) depends on the ratio of the diameters (particle over target) squared. The bigger the collector, the poorer the collection efficiency, with all other factors being the same. On the other hand, the bigger the particle to be collected, the greater the effect of the impaction and interception.

Diffusion is the particle moving under its own internal energy in random fashion. It may hit the target, and if it hits, we assume it sticks. It is not very significant for most scrubbing systems. As we look to the future where we must collect the smaller particles and condensed materials, it could become more important.

Even though we can go through these calculations for K_I and the other

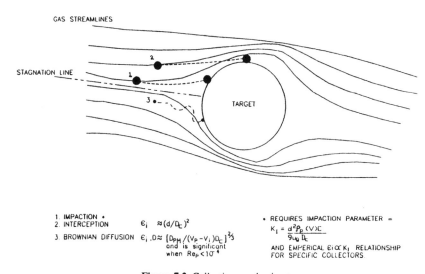

Figure 7.2 Collection mechanisms.

Figure 7.3 Diffusiophoretic force—negative.

collection parameters, we are not dealing with a single size target or a single size particle to be collected. There is a whole size distribution of both of these. It is interesting and important to recognize the relationships between these factors, but they are very difficult to use in real life.

Phoretic forces are weak, but sometimes important forces. They include diffusiophoresis, thermophoresis and photophoresis. For wet scrubbers, diffusiophoresis is the most important. To help explain this force, envision a hot, unsaturated flue gas that approaches a collector target (which is a droplet of water). Assume the flue gas is 300°F, and the water is fresh, so it is at ambient temperature. Figure 7.3 depicts this. In Figure 7.3, the little dots and circles represent dust particles that we are trying to get onto the collector target. We know it would help to increase the velocity differential between the two, but that is expensive. Is there another way? If we consider what happens in this situation, the gas will heat the water droplets. As the water becomes heated, it vaporizes and streams of water molecules flow away from the surface. This flue gas at 300°F is not saturated. Assume it contains typical combustion water, e.g., about 6 or 7% water. Therefore, the water vapor molecules easily flow away from the liquid water droplet. In doing so, they tend to oppose inertial impaction and actually push small submicron incoming particles away from the collector (big particles are not influenced). This action is dubbed negative phoretic force.

If we can have negative phoretic force, there must be a positive phoretic force. What can we do to make it positive? On the schematic in Figure 7.4,

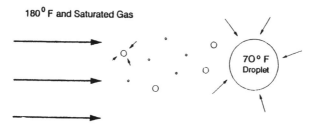

Figure 7.4 Diffusiophoretic force—positive.

the hot gas is shown cooled to 180°F and saturated. Now, as the gas approaches the 70°F liquid, it still will heat it up as before, but remember that as we heat the liquid, we cool the gas in our adiabatic scrubber. As the gas is saturated, there will be droplets of condensed liquid formed. There will be a tendency for the newly formed small condensation droplets to move toward the large collector droplet. The condensing water will also look for other nucleation sites. The smaller the site, the easier it is for condensation to occur. These condensing droplets will move onto the small particles, which then grow in size. So two things have been accomplished: we have created a movement of gas (water vapor) toward the water droplet, and we have also caused the small particles to grow. Both of these effects work to improve the collection efficiency. This situation is called positive phoretic force.

Other things that can be done to aid the collection of the particles is to charge them. It was found out that scrubbers following electrostatic precipitators work better. It's possible to do the same thing by adding a high intensity ionizer at the inlet of a scrubber to charge the particles. If we could get any of the additional attraction due to electrostatic forces, this would certainly be the benefit. This again is a very weak force. If you have an electrostatic precipitator preceding your scrubber, this could be beneficial. The scrubber will automatically take advantage of the benefit that the residual charge would give. Normally, on a new scrubbing installation, you design the scrubber to do the job without charging the particles. Later, if higher collection is required you can change the characteristic of the system by adding electrification.

7.8 VENTURI SCRUBBERS

There are various types of Venturi scrubbers as shown in Figure 7.5. There are Venturi scrubbers with movable throats with several different types of throats existing in the commercial market. What this does is keep the scrubber operating at a design condition. A relatively ideal throat velocity for a Venturi scrubber is about 150 ft/sec. You can achieve significant atomization as low as 70 ft/sec; but you lose effectiveness because inertial impaction has dropped off, and the atomization is not as complete as it is at 150 ft/sec. So, if you have a system that has a variable gas flow, what can you do to maintain desired conditions? One thing you can do is change the throat opening (area). As gas (volumetric) flow rate drops, close off on the throat by adjusting the dampers or moving the plumb bob up and down to close and open the throat cross-sectional area.

Venturi scrubbers work with cocurrent flow of gas and liquid. Venturi scrubbers are good particle collectors but are relatively poor gas ab-

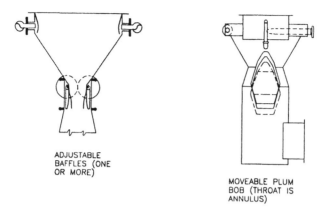

Figure 7.5 Venturi scrubbers with variable throat area.

sorbers. This is why some systems contain a Venturi for particles, followed by a packed tower for gases.

7.9 COUNTERCURRENT AND CROSS-FLOW SCRUBBERS

Packed towers are good scrubbers. They usually have countercurrent flow of gas and liquid. Spray towers are countercurrent for the most part but some are cocurrent. The cross-flow scrubbing unit has gases going perpendicular (horizontal) to the flow of the liquid. In this system, gas flows horizontally through a section filled with packing. Liquid enters and is distributed along the top, then flows vertically down through the system. Normally, you do not use packed towers for removal of particulates, as well as gases. However, the cross-flow system does function well with particulates. Normal countercurrent packed towers will plug if you introduce particulate matter into the bottom of the tower. Countercurrent and cocurrent spray towers are good for both particles and gases.

7.10 WASTE ENERGY POWERED SCRUBBERS

If you have waste energy (e.g., excess steam) in a system, you have ejector capability for powering the scrubber. This replaces the fan. Figure 7.6 shows steam ejector. The polluted gases enter from the top. The driving force, which in this system is the steam, aspirates and draws the gas through the system. In Venturi scrubbers, we introduced gas at a high velocity toward a liquid droplet target at a low velocity. The ejector is the

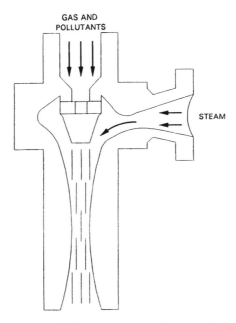

Figure 7.6 Ejector scrubber (no fan required).

reverse of this system, in that the steam with condensed liquid droplets is moving at the higher velocity and striking targets (i.e., striking the particles), which are moving at lower velocity. This could be likened to atmospheric rain or a spray tower, where the action is scavenging to remove the particles.

7.11 UNIQUE SCRUBBERS

The Calvert Collision Scrubber is a new concept and is shown in Figure 7.7. Gases come in at the left end of this figure, split and go both directions, turn and go through a throat where liquid is injected. The liquid is atomized. Typical Venturi-type collection occurs at this point on both sides. Then, the opposing streams move toward each other, and a second collision occurs. At this point, possible interactions include particle-to-particle, particle-to-droplet, droplet-to-droplet and droplet-to-particle. All of these combinations are utilized to enhance the collection. The rest of this system is described in the figure. The collision scrubber is more efficient than a conventional Venturi scrubber.

The other type of unique scrubber is the Catenary Grid Scrubber. This is a wide open tube, with one or more pairs of catenary grids, as shown in Figure 7.8. A catenary is the shape a chain would take when held at the

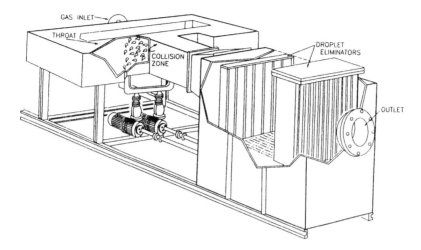

Figure 7.7 Calvert collision scrubber operating principle.

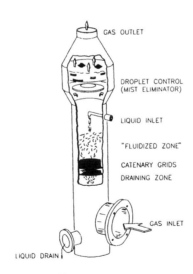

Figure 7.8 Catenary Grid Scrubber.

ends of the chain. The grid wire can be 1/4″ or 1/2″ opening wire, shaped into a catenary configuration and placed one above the other. Gases rise through the catenary and contact the liquid that is introduced from the top. Turbulent contact cells are formed that look very much like a fluidized bed. The cells have a total depth of about 18″, and they are in constant circulation through that 18″ zone. The complete mechanism of collection is not fully understood at this time. Centrifugal action creates sheets of water several microns thick, spaced about 200 microns apart in each cell. By having sheets of liquid in these cells, there is a tremendous amount of surface area. Surface area is needed for absorption, so a catenary grid scrubber provides good adsorption, as well as good particle control. There is a several second residence time in the active cell circular motion zone, so this scrubber gives good contact times.

The liquid leaves the bottom of the catenary. As it leaves, it agglomerates due to surface tension, which draws it across the catenary screen. The liquid agglomerates into huge droplets and falls out of the system. A Catenary Grid Scrubber has the capability of operating with just a few inches of water pressure drop. It can operate with very low liquid-to-gas ratios. These are operating cost advantages.

7.12 GENERAL INFORMATION

Each of these systems must be designed, maintained and operated properly. If they are not operated properly, they will not function as designed. In this chapter, we noted that particles approach and are captured by droplet targets. However, we must remember that it is important to have the complete throat area of the Venturi (or complete column area of a spray tower) covered with droplets to achieve efficient control.

7.13 MIST ELIMINATION/SEPARATION

The Achilles heel of a scrubber is mist elimination. Perhaps you have heard the statement: "Wet scrubbers release more particulates than entered the system." This could well be true. If you are circulating liquid in a closed-loop scrubbing system, that liquid may have 15–20% solids by the weight. One drop of liquid escaping through mist eliminator can carry a tremendous amount of particulate matter. You have to prevent the entrainment or reentrainment of liquid using mist eliminators. Basically, this can be done cyclonically or gravitationally, but the best is by inertial impaction, which is by the use of zigzag or mesh pad eliminators. Types of eliminators, along with typical velocities and material of construction, are summarized in Table 7.5.

TABLE 7.5. **Entrainment Separation (Mist Elimination) System Data.**

Equipment types—any of these may be internal or external of the scrubber

Gravitational sedimentation—useful and can be recognized when it occurs, but these types of separators are seldom built for use with scrubbers.

Centrifugal deposition—Cyclone separators are commonly used for mist separation (Actually, inertial impaction is a form of centrifugal separation.) Baffles added to cyclone separates can enhance mist removal. Overall, mist removal efficiencies are "low."

Inertial impaction—Systems containing elements to force abrupt changes in gas flow direction which causes inertial impaction of the entrained particles upon the elements. These elements may be mesh (fibrous layers); packing of saddles, rings, or other material; banks of tubes; perforated trays; or louvered or zigzag baffles.

ESP or other mechanical—not usually used, but can be used.

Gas velocity ranges for well-designed separators are typically:

Zigzag with up flow gases and horizontal baffles	12–15 ft/sec
Zigzag with horizontal flow gases and vertical baffles	15–20 ft/sec
Cyclone inlet velocity	100–130 ft/sec
Knitted mesh with vertical gas flow	10–15 ft/sec
Tube bank with vertical gas flow	12–16 ft/sec
Tube bank with horizontal gas flow	18–23 ft/sec

Material of construction:

Metal—these are heavy and seldom used.
Plastics—easily damaged or heat warped, but light weight and strong enough for most purposes
FRP—best (good to 400°F)
Polyester with flake glass coating on steel—not good

This table gives some more rules of thumb as to what the gas flow rates should be. If in doubt, go with the vendors' rates. For any specific type, there are ranges of velocities that are acceptable to give 99% particle removal. If you go below the lower velocity, you don't get good droplet throw-out (inertial impaction removal of the droplet). The droplets remain entrained and go out the stack. If you go above the recommended velocity value, you reentrain. You throw out the liquid droplets, but this liquid is resuspended, stripped off the surface of the mist eliminator and reentrained. There is a narrow range of acceptable velocities that you can work within.

7.14 OPERATING CHARACTERISTICS

When dealing with scrubbers, we need to understand some of the characteristics of operation. For Venturi scrubbers, you can obtain both particle and gas removal, as noted previously. For example, one can achieve good particulate control, with hot gas removal as a "bonus." The time in the Venturi scrubber is very short. Even though you have adequate mass transfer coefficients, driving forces, and surface area, all directed toward giving you good mass transfer, if you don't have time for this all to occur, then it's not going to be very effective. It has been indicated you can operate with throat velocities as low as 70 ft/sec. If you're looking for more gas removal and less particle removal, you slow down the velocity toward the 70 ft/sec minimum to achieve better gas removal. If you're looking for better particle removal than gas removal, increase the velocity toward the 150 (or even higher) ft/sec. The Venturi will run lower in pressure drop at the lower velocities and higher at the higher velocities. Ideally, you want to operate at pressure drops of 5–10 inches. This is best achieved with a properly conditioned gas that was quenched before coming to the scrubber. The quenching takes time and a little water to achieve saturation. It takes time for the water to evaporate, travel, and saturate the gas and pollutants. So, quenched conditions would result in more effective collection at lower pressure drops in the Venturi.

Table 7.6 gives typical operating characteristics for Venturi scrubbers. Note that typical liquid-to-gas ratios (L) for absorption are much higher than for particle removal. These are just suggested averages; some systems operate much higher than 40 gal/1000 ft³ and may go all the way up to 120. For gas absorption, you are trying to provide surface area, and for particle capture, we are simply providing the targets. The lower L is a cutoff value (i.e., if you don't have at least 5 gallons per 1,000 ft³ in a Venturi, you don't

TABLE 7.6. Operating Characteristics of Venturi Scrubbers.

Pollutant	Pressure Drop	Liquid-to-Gas Ratio	Liquid-Inlet Pressure	Removal Efficiency % for Gases and Cut Diameter for Particles*
Gases	5–100 in. of Water	20–24 gal/ 1000 ft³	1–15 psig desirable; up to 25 psig typical	30–60% per Venturi Depending on pollutant solubility and operation 90–99% is typical for staged systems
Particles	10–100 in. of water 20–60 in. of water is common	5.0–20.0 gal/ 1000 ft³		0.2 μm cut diameter depending on ΔP

*Cut diameter is the size particle collected with 50% efficiency.

have enough droplets to cover the throat). Again, this is one of those "magic" numbers that indicate a scrubber is in trouble—if you find you have less than five for the liquid-to-gas ratio.

Ideally, you want to keep spray nozzle pressure low, to save pumping costs and to prevent creating excessively small droplets. It depends on the nozzle, liquid rate and type, temperature, particle size to be captured, and the droplet size desired. Mist eliminators need to be cleaned with liquid spray, usually as a freshwater spray to the bottom of the mist eliminator as needed. That pressure should be low to keep from blasting the dirt right through it.

Venturi pressure drop is a function of several parameters. Pressure drop in a Venturi can be increased by increasing the throat velocity, increasing the density of the gas, increasing the area of the throat and increasing the liquid-to-gas ratio, according to the Hesketh equation:

$$\text{Venturi } \Delta P = \frac{V_T^2 \varrho_g A^{0.133} L^{0.78}}{1270}$$

An "optimal" Venturi throat length can also be estimated using another Hesketh equation:

$$X_T = 328.582 V_T^{(0.02343L - 0.8657)} \exp(-0.63L)$$

where:

ΔP = pressure drop, inches water
V_T = throat velocity, ft/sec
ϱ_g = gas density, lb/ft³
A = throat area, ft²
L = liquid-to-gas ratio, gal/1000 acf
X_T = throat length, inches

Some Venturi are built that have no throat, only a converger and diverger. This doesn't give time for adequate collection. It takes time for the particles to be accelerated and to reach the target droplets. On the other hand, if you make the throat too long, particles may not approach the targets, as both of them will be moving at the same velocity. That is, the droplet collector will be accelerated to the gas velocity also, and the two will travel as a pair down the throat.

Pressure drop across the working portion of a wet scrubber is important to cost of operation but is relative to particle collection efficiency of the device. This is shown for several wet scrubbers in Figure 7.9. Note that a lower cut diameter is a more efficient scrubber. In general, as scrubber

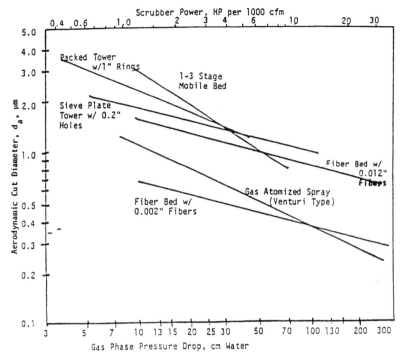

Note: Packed Tower, Mobile Bed and Sieve Plate Tower are counter current flow
scrubbers. Fiber Bed are co-current flow scrubbers.

Figure 7.9 Approximate energy-efficiency characteristic for types of wet scrubbers.

pressure drop increases, efficiency of particle collection increases. The highly efficient Calvert flux force condensation scrubbers (quench plus Venturi) can remove metals as noted in Table 7.7. Note that this includes condensing the vapors and collecting the particles.

7.15 WET SCRUBBING SYSTEMS

An example of a typical "complete" system is given in Figure 7.10. Many tasks, pumps and liquid handling portions are required. For systems such as these, a summary list of operational items includes the following considerations.

(1) Most important parameters in order of importance are
   - ΔP
   - liquid-to-gas ratio (L/G, or simply L)
   - gas flow rate (or liquid flow rate)
(2) For a constant size scrubber with constant L/G, a decreasing ΔP may mean

TABLE 7.7. Flux Force Condensation Scrubbing System Heavy Metal Removal Data.

Metal Compound	Removal Efficiency*
Arsenic	99 + %
Beryllium	99%
Cadmium	99 + %
Chromium	99 + %
Copper	95 + %
Iron	97 + %
Lead	98 + %
Mercury	98 + %
Nickel	88 + %
Vanadium	99 + %
Zinc	90 + %

*Based on incineration of hazardous liquid waste.
Courtesy: Calvert Environmental.

- fan failure
- leaks
- erosion
- downstream pluggage
- decreasing gas and/or liquid rates

(3) An increasing ΔP may mean
  - blower problems
  - solids build up in scrubber or mist eliminator pluggage within the scrubber (usually increasing ΔP signifies increasing gas and/or liquid rates)

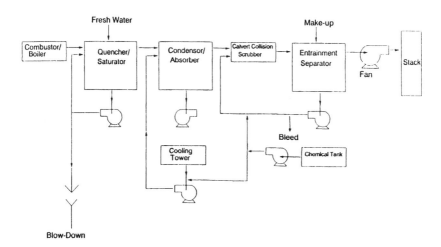

Figure 7.10 Example of a "complete" scrubbing system.

(4) Record data for
  - scrubber ΔP
  - eliminator ΔP
  - pHs wash rates of frequency
  - recycle solids rate and concentration
  - feed rate
  - bleed rate
  - inlet and outlet rates
  - nozzle pressure
  - temperatures
  - O_2 and/or CO concentration in flue gas
  - solids ratio and % H_2O
  - slurry retention time and oxidation rate

(5) Mist eliminators should operate at >30% design capacity.

(6) Temperatures and pHs cannot exceed material of construction limits:
  - rubbers 175°F
  - fiberglass reinforced plastic 190°F
  - lead 250°F

(7) A typical suggested scrubbing maintenance checklist includes:

Daily Inspection

Check Point	What to Look For
1. Pump	a. Leaking at gland
	b. Increased noise
2. Valves	a. Position
	b. Leaks
3. Piping	a. Leaks
4. Body	a. Leakage
5. Pressure gauge	a. Pressure change from previous day
6. Pressure gauge ammeter combination	a. Changes in either or both pressure reading and ampere draw from last clean system check readings

Gauge Readings	Probable Trouble
1. a) Water pressure same	None
b) Ampere draw same	a. Missing nozzles; pump wear or plugging of suction line
2. a) Water pressure decrease	
b) Amperage draw decrease	b. Ditto

3. a) Water pressure same or increase   a. Plugging vs. wear in nozzles or spray bars
    b) Ampere draw decrease   b. Plugging vs. wear in nozzles or spray bars
4. a) Water pressure increase   a. Holes in spray bar or manifold
    b) Ampere draw increase   b. Holes in spray bar or manifold

Weekly Inspection

Check Point	What to Look For
1. Spray bars	a. Plugged nozzles
	b. Worn or missing nozzles
2. Pipes and manifolds	a. Plugging or leaks
3. Pressure gauge	a. Check accuracy
4. Pumps and valves	a. Wer
	b. Valve operation
	c. Corrosion

CHAPTER 8

Principles of Dry Scrubbers

FRANK L. CROSS, JR.

8.1 INTRODUCTION

A dry scrubber causes acid gases (e.g., SO_2, HCl, HF, etc.) to react with an alkaline material (i.e., lime). That forms a particulate that is removed from the flue gas stream by a particulate control device, which is usually a filter (baghouse).

8.2 CHEMISTRY FOR ACID GAS CONTROL

A dry scrubber system typically consists of two main units, a spray or injection reactor, followed by a baghouse. The two variations of this system, which are commonly used, differ in the way the reagent is injected. Spray dryer systems inject the reagent as a wet slurry while dry sorbent injection systems use a dry powder. Spray dryer systems have traditionally been used in conjunction with large-scale industrial processes. Dry sorbent injection is now emerging as a control technology particularly suited for small- to medium-sized industrial boilers.

In a dry system, the flue gas leaves the waste heat boiler and then enters a cooling device such as a quench chamber. In the quench chamber, the gas is cooled as a conditioning step to allow the use of a baghouse.

The conditioned flue gas then enters a reactor vessel where the reagent, a dry powder, is pneumatically injected. The reagent injection takes place in a high velocity mixing zone to maximize the chemical reaction. Neutralization of the acid gases occurs as the reagent reacts with the acid to form reaction salts and water. Typical reactions are represented as:

119

$$Ca(OH)_2 + HCl \rightarrow CaCl_2 + 2H_2O$$
$$Ca(OH)_2 + SO \rightarrow CaSO_3 + H_2O$$

The gas is then psssed to a baghouse or electrostatic precipitator. Baghouses have generally shown higher removal efficiencies than precipitators. In the baghouse, fabric filters consisting of cylindrical filter bags are vertically suspended in a baghouse compartment. The filter bags are porous, which allows the gas to pass through them while capturing a high percentage of the particles suspended in the gas stream. The fly ash, reaction salts and unreacted reagent are captured in the baghouse as the solids collected on the filter bags. Further HCl, SO_2 and acid gas removal takes place on the filter bags as the flue gas passes through the filter cake containing unreacted reagent. As the particulate cake builds on the bags, the pressure drop across the bags also increases and the bags must be cleaned. Cleaning the bags dislodges the particulate cake, which falls into a hopper for removal and disposal.

Dry sorbent injection followed by a baghouse has a number of advantages, as well as disadvantages. There is no liquid discharge from the system, but the residue collected in the baghouse must be disposed. The water used in the quench chamber is completely evaporated so there is no worry about a visible steam plume coming from the exhaust stack.

8.3 DRY SCRUBBERS

8.3.1 CONFIGURATION

The general arrangement of a dry scrubber includes a cooling device(s)

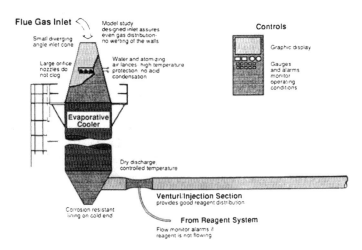

Figure 8.1.

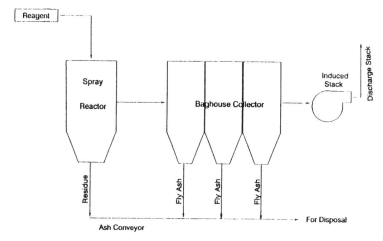

Figure 8.2 Dry scrubber system.

followed by a reagent injection system, followed by a baghouse (including fan and stack). The cooling device(s) may include a boiler and/or evaporative cooler (or similar device) (see Figure 8.1).

The wet/dry scrubber (see Figure 8.2) uses an alkaline liquid atomized for the acid gas reaction to form a particulate. The dry/dry scrubber (see Figure 8.3) uses a dry, powdered, alkaline material (i.e., lime) for the acid gas reaction. Notice that the flow diagram for the dry/dry scrubber requires chemical storage, ejectors, blowers, etc.

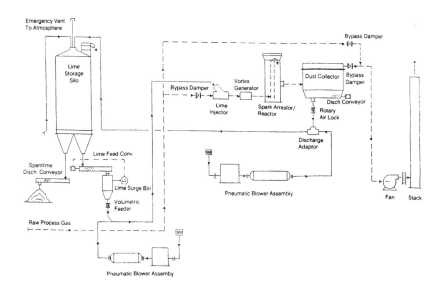

Figure 8.3 Dry/dry scrubber.

8.3.2 RESULTS OF DRY SCRUBBERS

The recent interest in dry scrubbers has been to demonstrate good control results for furans and dioxins (see Chapter 3, Table 3.6). In addition to good results, the dry scrubber baghouses usually have excess alkaline material on the bags, which acts as another collector for excess concentrations of acid gases from fluctuating gas flows.

8.3.3 RESULTS OF SMALLER FACILITIES

The strenuous emissions requirements for boilers such as medical waste incinerators (see Table 8.1) for New York State were based upon dry scrubbers for APC (Air Pollution Control). The initial dry scrubber data (i.e., Erlanger Hospital, Chattanooga, TN) was of concern, but now some of the smaller dry scrubbers are performing satisfactorily (see Table 8.2).

8.4 RETROFIT FOR BOILER SYSTEMS

The following is a typical concept retrofit report for a smaller industrial user.

8.4.1 INTRODUCTION

_____ has received several enforcement notices concerning visible emissions from its coal-fired boiler. To correct the problem, _____ wishes a turnkey installation of a baghouse type collector at this time, with possible conversion to a dry scrubber in the future, should the use of compliance coal become a problem.

_____ was asked to provide a preliminary concept report of the retrofit for budget and scheduling purposes prior to the receipt of a formal turnkey proposal from _____. The concept report is to address location, size and cost of the proposed air pollution control system.

8.4.2 EXISTING SYSTEM

The existing coal-fired boiler is a three-drum sterling, water-tube B&W boiler that was built in 1947. It is equipped with a Detroit Stoker roto-grate and has an air preheater. The rated steam capacity is 40,000 lb/hr of superheated steam.

Particulate emissions from this boiler are controlled by a Western Pre-

TABLE 8.1. Infectious Waste Incinerator Regulations.

State/Parameter	Indiana	New York	Penn.	Alabama	Minn.	Miss.	Calif.	Wisc.
Air Emissions								
Particulates	0.3 lb/100 lb DG	.015 gr/dscf at 7% O_2	.03 gr/dscf at 7% O_2	.02 lb/100 lb feed	.01 gr/dscf	.2 gr/dscf at 12% CO_2	0.1 gr/dscf	.03 gr/dscf at 12% CO_2
Visible emissions (opacity)	—	Hourly avg. 10% max. continuous 6 min. avg. <20%	30%	20%	—	40%	20%	5% (as measured by USEPA Method 9)
HCl (acid gas)	—	Less restrictive 90% HCl reduction or 50 ppm HCl	30 ppm	—	Testing required	—	—	50 ppm at 12%
SO_2	—	50 ppm	30 ppm	—	—	—	—	CO_2 over any continuous 1-hr period

TABLE 8.2. Results of Tests from Dry Scrubbers on Hospital Incinerators.

Company	Location	Feed Rate	Particulate GR/DSCF, 7% O₂	HCl	HCl Removal Efficiency	NOₓ PPM	CO PPM	SO₂ PPM	THC PPM	APC Equipt.
Partyka Resource Management Company of Oklahoma	Stroud Oklahoma	90 TPD (7529 LB/HR)	0.0141 (Front Half, 12% CO₂)	43.6 PPM @ 7% O₂ 2.97 LB/HR	97.2	86	1.1	7	2.9	United McGill Dry Scrubber
Florida Hospital	Orlando	1200 LB/HR	0.0168	16.43 LB/HR (not a dry scrubber)			0.625			Lime Inject. with Baghouse United McGill
Medi-Waste LTD	West Babylon New York		0.0047	0.241 to 82.0 PPM @ 7% O₂ 0.605 LB/HR	90					Interel Dry Scrubber
Fairfax Hospital	Fairfax Virginia	825 LB/HR	0.0086 0.38 LB/HR	28 PPM	96.6					Consumat Dry Scrubber
Erlanger Medical Center	Chattanooga TN	870 LB/HR	0.020 @ 12% CO₂	438 PPM 11.42 LB/HR	50	89.9				Boet Dry Scrubber

cipitation mechanical collector. Cleaned flue gas is exhausted to the atmosphere via a lined 40-foot high stack that has a diameter of 4 feet.

The boiler is fired with low sulfur compliance coal from the _____ mine.

8.4.3 RETROFIT SYSTEM

8.4.3.1 Concept

The initial concept is to install a baghouse to solve the immediate visible emissions problem (Phase I) and eventually add a lime injector and convert the system to a dry scrubber (Phase II).

8.4.3.2 System Design

The system being proposed is illustrated in Figures 8.4 and 8.5. These sketches illustrate the installation of a baghouse elevated above the floor on a grate supported by beams at the back of the coal-fired boiler. The 4.5-foot duct would be interrupted with a stop panel. (Note: This is actually a damper that can be used as a bypass in case of an emergency.) The exhaust gases would then be drawn through the baghouse by the exhaust fan on the roof. The mechanical collectors would remain in the system to act as precleaners and spark arrestors for the baghouse. The fly ash collected in the baghouse would be removed with a screw conveyor and rotary valve arrangement. The discharge pipe from the rotary valve would connect by gravity to the existing ash removal systems for the boiler.

The lime storage silo and injector for Phase II could be installed now or at a later date.

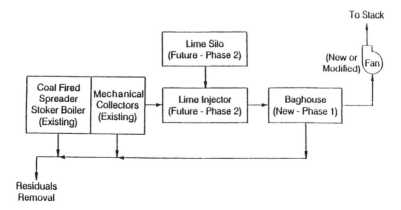

Figure 8.4 Conceptual flow diagram of proposed system.

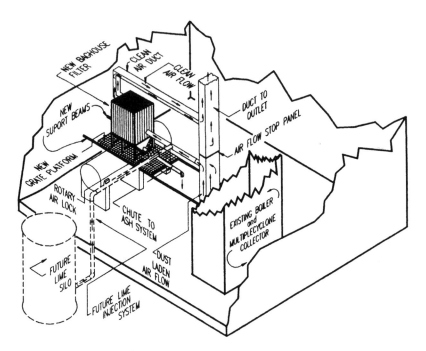

Figure 8.5 Emission control for existing coal-fired boiler.

8.4.3.3 Baghouse Criteria

The baghouse specifications are as follows:

- A/C ratio @ 4:1
- size of bags: 6-9/16″ dia. × 12′ long
- number of bags: @ 372
- cleaning: pulse jet
- type of bags: ryton material
- height of collector: 32′
- approximate foot print of collector: 12′ wide × 24′ long
- gas flow: 29,613 acfm
- inlet gas temperature: 370°F
- Inlet dust loading: _____
- Collector weight @ 32,000 lbs

8.4.4 SYSTEM COST

The estimated cost in 1991 year dollars for this retrofit conversion of the stoker-fired boiler at _____ is as follows:

Phase I – Baghouse	Cost ($)
Equipment cost	147,982
Installation (45%)	66,591
Engineering (10%)	14,798
Permitting (3%)	4,439
Testing (3%)	4,439
Contingencies (7%)	10,359
TOTAL	$248,608

(Percentages are of equipment cost.)

Phase II – Dry Scrubber	Cost ($)
Equipment cost	135,000
Installation (45%)	60,750
Engineering (10%)	13,500
Permitting (3%)	4,050
Contingencies (7%)	9,450
TOTAL	$222,750

(Percentages are of equipment cost.)

TOTAL PROJECT (PHASE I & II)	$471,368

These numbers (i.e., erection costs) do not include wall and/or roof removal costs to allow construction access.

8.4.5 TIME TABLE FOR THE PROJECT

Figure 8.6 is a Milestone Chart for the project indicating completion of the project (either both phases or Phase I) by June of 1991. This schedule may be influenced by regulatory action, company decision or manufacturer/installation considerations.

8.5 SPECIFICATIONS

This section provides an example of a typical specification of a dry scrubber system for air pollution control. The system to be specified is shown in Figure 8.7.

Operating parameters and supporting calculations are as follows:

- volumetric flow rate: 29,613 acfm
- assume air/cloth ratio of 4:1

Cloth Area Required

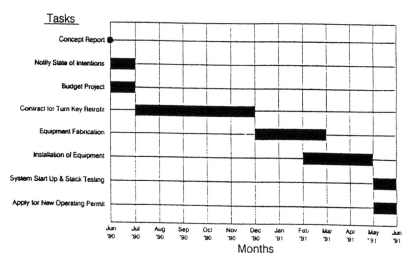

Figure 8.6 Milestone chart for retrofit of coal-fired boiler.

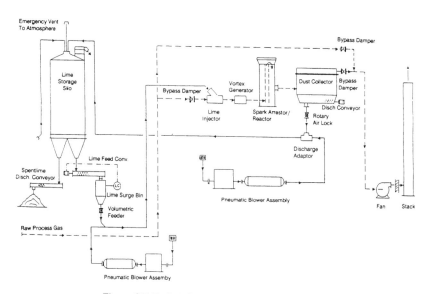

Figure 8.7 Typical flow diagram for a dry scrubber.

$$A(\text{ft}^2) = \frac{Q(\text{acfm})}{V(\text{ft/min})}$$

$$A = 29{,}613/4 = 7403 \text{ ft}^2$$

- closest commercial size = 7596 ft²
- ∴ A/C ratio = 29,613/7596 = 3.9:1
- estimated bag cost: 7597 ft² × $2.90/ft² = $22,028

Use $22,000

The following eleven items discuss the specifications for the dry scrubbing air pollution control system. This is intended as an example system, and an infectious waste incinerator is the example type of boiler system.

8.5.1 GENERAL

(1) The CONTRACTOR shall provide a dry chemical scrubber air pollution control system specifically designed for control of particulate and acid gas emissions from an infectious waste incinerator.

8.5.2 HYDRATED LIME STORAGE

(1) A single storage silo shall be provided to serve the air pollution control system and sized for seven days storage capacity, 6 ft. max. diameter, single compartment. Roof shall be designed to support all roof mounted equipment plus a superimposed live load of 50 lbs per square foot. Silo shall have 70 degrees cone bottom. Silo shall be furnished with connections in the straight section for low level bin signal connection and high level bin signal.

(2) Roof shall be furnished with a 20 in. diameter combination manhole inspection port, including dust-tight closure with locking hasp, pressure/vacuum relief valve, dust filter adapter and air-filled turbulence discharge box.

(3) Storage section shall be fabricated of minimum 3/16 in. plate and equipment section shall be fabricated of minimum 1/4 in. plate. All steel plates shall be ASTM A36. Bottom flange shall include three leveling screws and anchor bolt saddles.

(4) Equipment section underneath shall be furnished with two 36 × 84 in. access doors with hardware and two vents with bird screens.

(5) Complete storage and equipment section shall be of welded construction, shipped in one piece, and furnished with lifting lugs.

(6) The storage silo shall be furnished with an external ladder complete

with cage, platforms at bin level indicator and at intervals not exceeding 30 ft., and steel support members. Platforms shall be provided with handrails and toe plates. Handrails and toe plates shall be provided around the perimeter of the silo roof.

(7) The silo discharge will have a motorized bin activator (5 ft. min. dia.) to prevent bridging, jamming, and segregation and to ensure positive flow of material on a first-in, first-out basis. Design of the bin activator discharger shall prevent transmission of vibrations to the storage silo. Bin activator shall be of the following materials of construction:

- contact materials – carbon steel
- external support brackets – carbon steel
- flexible screws – suitable reinforced natural or man-made elastomer
- motor – 1.5 HP, TENV, 460 volt, 60 hertz, 3 phase
- outlet size – shall be determined by the manufacturer, stated in the bid and approved by the ENGINEER

(8) Two level indicators shall be furnished, one for high level and one for low level. Indicators shall be motor operated type with stainless steel paddle or vibrating probe. High level indicator shall be furnished with 3 in. shaft extension on silo roof. Low level indicator shall be furnished with paddle guard adjacent ladder in straight walls of silo. Indicators shall be accessible from the ladder and premounted prior to shipment.

(9) Storage silo fill pipe shall be 4 in. dia., schedule forty seamless steel furnished complete with 4 ft radius elbows, flanges, truck connection, dust cap with chain, and limit switch on end of pipe to provide automatic operation of the silo vent filter during and after truck unloading cycle. Limit switch assembly shall be adjustable and secured to a heavy-duty mounting bracket. Limit switch shall be complete with one set of N.O. and one set of N.C. contact. Spare contacts shall be wired into the unloading control panel to numbered terminal strips for remote monitoring of filling operation. Fill pipe shall extend through outside plant wall and shall terminate approximately 5 ft., 0 in. above grade at the truck unloading/silo fill station and shall be shipped unassembled for field installation.

(10) Silo shall be furnished with a roof mounted dust filter with retained dust to discharge directly into storage bin. The baghouse shall have a suitable cloth area to handle 1200 scfm. The baghouse assembly shall be in a steel housing and weatherproofed with gasketed access doors.

(11) The storage silo shall be furnished with a truck fill panel mounted

near the truck unloading/silo fill station. Truck fill panel shall be in a NEMA 4 enclosure. Panel shall include

- selector switch, key operated, for power ON/OFF
- push button for manual dust filter pulse operation
- push button for alarm silence
- indicating lights for: power "ON," dust filter pulse "ON," high bin signal, low bin signal
- alarm horn

8.5.3 LIME FEEDER

Lime feeder shall be volumetric dry solids type screw feeder with adjustable capacity for the range required and suitable for handling hydrated lime with a fineness varying from 75% passing a #200 sieve to 99% passing a #325 sieve. Flow into the feeder shall be through a 3 cu ft feed hopper equipped with a vibrator. Feeder shall be complete with a full wave SCR controlled with 4 to 20 maDC proportional pacing DC variable speed drive having a 30:1 output range. All speed adjustments shall be stepless and can be made with the feeder operating. Controller shall be in a watertight, dust-tight NEMA 4 enclosure mounted adjacent to the main control panel, and shall include a digital speed selector, ON-OFF switch, line fuse and armature fuse. The drive motor shall be TEFC, 1800 RPM direct coupled to 30:1 right-angle gear reducer.

8.5.4 LIME TRANSFER

Depending upon system arrangement, the lime feed can either be gravity flow into the duct or pneumatically conveyed.

(1) A motor operated rotary vane feeder will be provided to form an airlock. It will be designed to be leakproof at up to 150% of the pressure differential expected.

(2) Gravity feed: Where feasible to mount the lime silo above the duct injection point, lime may be fed gravimetrically into the duct.

(3) In lieu of gravity feed, provide a pneumatic dry sorbent transfer system including a positive displacement blower, 3 in. sch 80 conveying pipes, long radius elbows, fittings, compilings and pipe supports.

8.5.5 HEAT EXCHANGER

A heat exchanger shall be provided to cool flue gases from the waste heat recovery steam generator to enhance the acid gas reaction. Preferred design type is reverse shell-and-tube; flue gas on outside of cooling tubes, cooling air blown through insides of tubes.

(1) Cooling air shall be automatically controlled to maintain outlet flue glass temperature of a minimum of 30°F above the flue gas saturation temperature.

(2) Maximum allowable gas outlet temperature variation is plus or minus 10°F from set point.

(3) Maximum allowable pressure drop across heat exchanger at maximum charging rate operating conditions is 3 in. wg.

(4) Heat exchanger shall be equipped with soot blowers and/or sonic horns for automatic cleaning of cooling tubes at adjustable time intervals.

(5) Horizontally mounted, replaceable cooling tubes with Viton seals are to be provided.

(6) Gas inlet is to be at the top and the gas outlet is to be at the bottom of the unit.

(7) In the bid, give special consideration to the difficult nature of this application and state method of protecting against condensation and associated operational problems. State tube material and shell construction and lining.

8.5.6 REAGENT INJECTION AND REACTION CHAMBER

Hydrated lime shall be injected through a nozzle designed to evenly disperse the hydrated lime across the flue gas stream. A variable throat Venturi will be provided upstream of the nozzle to minimize fallout during low load conditions.

(1) Lime injection will be between the heat exchanger and a reaction chamber.

(2) CONTRACTOR is to supply a reaction chamber if required to get suitable reaction time for efficient reagent usage. Construction is to be minimum 3/16 in. plate and complete with hopper for any fallout. Provide in bid an analysis demonstrating that design provides sufficient reaction time and state reagent usage rate used as a design basis.

8.5.7 FABRIC FILTER

The baghouse shall be an outside bag collection type with pulse jet cleaning and shall consist of a complete, fully assembled module and shall be shipped with hopper attached.

(1) Housing and tube sheet: The housing and tube sheets shall be a minimum of 1/16 in. thick and of all welded construction. Tube sheets shall be seal welded to the module walls and shall include stiffeners to limit the deflection of the tube sheet under loading due to the bags, cages, Venturis, dust load and foot traffic to a maximum of 0.015 in. per foot. The arrangement shall be such as to permit individual removal of bags and cages from the top of the fabric filter.

(2) A dust hopper shall be included and shall be sized for a minimum of four hours storage time based on 30 pounds per cubic foot ash density. Structural design and loadings shall be based on a density of 100 pounds per cubic foot based on the hopper full of ash. The hopper shall have a minimum valley angle of 55 degrees. It shall be a pyramidal type and shall be constructed of 3/16 in. minimum carbon steel plate.

- The CONTRACTOR shall furnish one, six-inch square stike plate on the hopper. Space shall be provided on the hopper for future vibrators. One, four-inch-diameter rod-out port is to be provided.
- The CONTRACTOR shall provide one vibrating type level detector for indication of high hopper dust level.

(3) Bags and cages
- Bags and cages will be cylindrical, nominally 6 in. dia. and mounted vertically, and not to exceed 12 ft in length.
- Bag cages shall be fabricated in one piece. Split cages are not acceptable. Each cage shall be constructed of wire mesh or rigid wires to provide suitable bag support. The cage will have a solid bottom pan and an integral Venturi or nozzle at the top. All cage materials shall be galvanized steel. Finished cage shall be of rigid construction and shall be cylindrically smooth and straight throughout its full length. Special care shall be taken to ensure there are no rough spots on cages to cause bag abrasion.
- Cage attachment to tube sheet shall be rigid to minimize movement of bags during operation.
- Bag materials shall be polyimide (trade name P-84) and should have a maximum air-to-cloth ratio of 5:1.
- Bag and cage design shall be such that bag length will be fitted to cage length and no fabric folds will exist on installed bags. CONTRACTOR shall describe design used to accomplish this. Any exposed bands or clamps used shall be stainless steel. A double thickness wear strip shall be provided at the bottom of the bags to resist abrasion from bags hitting each other.
- Bag life shall be guaranteed for a minimum of two years.

(4) Bag cleaning equipment
- The baghouse shall have a factory installed compressed air manifold. The manifold shall consist of a steel pipe or square tube, mounted external to and running the full length of the unit. The manifold shall include a drain port.
- Between the air manifold and blow pipes for each row of bags shall be a double diaphragm type air pulse valve. Each air pulse valve shall be controlled by a direct connected 110 volt, single phase, 60 hertz, solenoid pilot valve. Remote solenoid valves connected by tubing will not be permitted. Solenoid valves shall be prewired to a common junction box.
- Blow pipes shall be provided above rows of bags. A maximum of sixteen bags may be pulsed from each flow pipe and air pulse valve. Blow pipes shall be fabricated of Schedule 40 steel pipe with orifice holes located accurately over Venturis. Blow pipe system shall be designed and rigidly supported at each end to maintain accurate lateral and vertical orifice hole/Venturi alignment and spacing. Valves, piping, wiring, etc., shall be protected from weather and traffic and are to be easily accessible for maintenance.

(5) Tube sheet bag and cage access
- CONTRACTOR shall provide preinsulated lift-off or hinged top door(s) for bag access. Door shall incorporate a 3 in. layer of thermal insulation between a checkerplate top suitable for foot traffic and a 3/16 in. steel plate on the underside. Doors shall be secured by hold down clamps and shall have high temperature seals to minimize flue gas leakage or air infiltration and provide a weatherproof seal with the module.

8.5.8 DUCT AND DAMPERS

(1) Provide ductwork interconnecting the heat exchanger, lime injection/reaction chamber, and fabric filter plus outlet duct from the fabric filter to an I.D. fan location.

(2) All duct is to be a minimum of 3/16 in. plate and sized for a nominal 360 fpm velocity at design conditions.

(3) The CONTRACTOR shall include an analysis or model study of duct and equipment to ensure good dispersion of reagent and minimum reagent/ash fallout.

(4) Damper valves:
- The module will have both an inlet and an outlet isolation valve. Both shall be pneumatically operated butterfly valves with provisions to manually lock them in the closed position. Each

damper shall be shop assembled and tested and shall be equipped with flanges for connection to the inlet and outlet. They shall be provided with limit switches to indicate both full open and full closed positions. Limit switches shall have NEMA 4 enclosures and DPDT contacts.

- A double-bladed poppet valve is to be provided for system bypass from the inlet to the outlet. There will be two (2) sealing plates to ensure 100% isolation. This valve will have properly stiffened stainless steel valve plate and stainless steel shaft. The valve is to operate in a vertical position. Damper shall be operated by a pneumatic actuator with provisions for locking the valve in position. Provisions shall be made for field alignment of the operator. Damper operating speed shall be completely adjustable during both opening and closing by means of a speed adjusting mechanism. Actuator shall be provided with flexible air hose(s), air filter and lubricator. Regulators shall be provided if required. The damper shall be provided with limit switches to indicate both full open and full closed positions. Limit switches shall have NEMA 4 enclosures and DPDT contacts.

(5) A bypass duct connection between the inlet and outlet will be provided.

8.5.9 SUPPORT STEEL AND PLATFORMS

(1) The CONTRACTOR shall provide supports for all provided equipment. The support steel should have a level tolerance of plus or minus 1/8 in. The structural support steel shall be plumb, square, and aligned in accordance with AISC Chapter 5, Section 7.11.
- The fabric filter will have supports to provide for a clearance of 5′, 6″ below the hopper flange.
- If the heat exchanger or reaction chamber has a hopper for ash fallout, provide supports for a clearance of 5′–6′ below the hopper flange.

(2) Access facilities shall be provided to permit access to items of equipment that require maintenance, servicing, or inspection. Access doors shall be provided for internal access to the baghouse and components described below.
- Service platforms shall be provided at any level for access to all module dampers, operators, bag cleaning systems, and access doors. Valves, piping, wiring, and the like, shall be easily accessible for maintenance from an adequate walkway.
- Service platforms shall also be supplied for any equipment that

may require routine maintenance including heat exchanger fan, lime injection equipment, lime silo accessories, etc., unless equipment is sufficiently near the ground and infrequently serviced to make it practical to use temporary ladders. Provide in bid location of all service platform included with the system.

- Access to all service walkways from grade shall consist of caged ladder(s).
- Walkways, ladders, stairways, grating and handrails are to be designed per federal and state OSHA requirements.

8.5.10 AUXILIARY HEATING AND INSULATION

The CONTRACTOR shall provide an auxiliary heater and insulation designed to maintain a minimum system operating temperature of 20°F above the saturation temperature of the gas stream. The CONTRACTOR shall include in the proposal a design analysis to show that the equipment proposed will meet this performance criterion.

8.5.11 CONTROL SYSTEM AND INSTRUMENTATION

(1) A complete integrated system of instruments and controls shall be provided to measure, monitor, operate and control the system. The air pollution control system shall be operated automatically from a main control panel located near the baghouse. This panel shall be complete and shall house all instruments, controllers, indicators, relays, power supplies, fuses, indicating lights, push buttons, switches, transformers, contactors, programmable controller, alarm contact, and other devices as required to provide a complete and operable system consistent with the intended operation as described herein.

(2) The CONTRACTOR shall provide all required local instruments, transmitters, sensing probes.

(3) Instrumentation power shall be 120 VAC, 1 phase, 60 Hz.

(4) Instrumentation control signal shall be 4–20 m.a. DC.

(5) Instrumentation enclosures shall be NEMA 3R.

(6) Switches and contacts shall be DPDT with dry contact to operate at 24 VDC or 120 VAC.

(7) All electrical equipment components such as, but not limited to, switches, contacts, relays, enclosures, powered instruments, etc., shall be listed by Underwriters Laboratories (UL). If required to do so, the CONTRACTOR shall provide the Company proof of UL listing of a component(s) in the form of a photocopy of the UL guide card for the component(s) in question.

(8) Cabinet structure
- Panels shall be NEMA 3R, freestanding or wall mounted design for front access and fully enclosed construction. Louvers, exhaust fans, and heaters shall be provided, if required, to maintain electronic instrumentation within CONTRACTOR's specified operating temperature range.
- All internal panel electrical components shall be open type (not enclosed). All panel wiring shall be open type, neatly bundled and supported, or semi-open, run in plastic wiring troughs, and shall be provided with unique wire number markers (plastic sleeve type). All wiring requiring external connections to the panel shall be run to terminals properly marked as per the approved shop drawings for external connections. Hand-lettered or adhesive type terminal identification is not acceptable. Not more than two field connected wires shall be terminated on any one point.

(9) The system shall be operated, monitored, and controlled by a single main microprocessor-based programmable logic controller (PLC), located in the control panel. Push buttons, selector switches, and field device signals from limit switches, timers and differential pressure transmitter shall be handled as inputs to the PLC logic, which shall control all necessary interlocks, operating logic, signal interpretation, and field devices. The control system shall monitor and operate the heat exchanger, lime injection and baghouse cleaning.
- Heat exchanger controls: The control panel will modulate flow of cooling fluid to the heat exchanger to maintain the set point outlet temperature.
- Lime injection: There will be a manual adjustment for rate of lime injection. In addition, an alternate automatic mode is to be available that will accept a customer supplied 4–20 m.a. signal for adjustment of flow.
- Baghouse system cleaning shall be initiated by pressure differential across the system. There shall be an adjustable timer for secondary initiation of cleaning. Once cleaning has started, it will continue until the pressure drop across the bags reaches an acceptable predetermined value. The following features and functions shall be incorporated into the system.
  a. Cleaning times (duration and interval) shall be adjustable.
  b. Differential indicator/switch for baghouse differential pressure from inlet to outlet shall be provided. Inputs from differential switch shall be used by PLC for control of module cleaning operations as required.
  c. Inlet, outlet, and bypass damper position switches shall be used

in the control logic for damper position failure and to indicate when the system is off-line for servicing.

    d. Chart recorder: One (1) two pen, 12 in. dia. circular linear chart, internally illuminated, hinged front, disposable filter type pans. Include 400 charts of 24-hour duration and two (2) spare sets of pens. Recorder shall record inlet gas temperature and fabric filter differential pressure.

    e. Baghouse inlet RTD with well-mounted inlet manifold complete with temperature transmitter if required. Temperature high switch shall also be provided. The RTD is to be 100 OHM platinum type (at 0°C) with 304 stainless steel sheath, standard three-wire lead, and weatherproof connector lead.

    f. Compressed air low pressure switch with separate dry contacts for remote sensing.

- Graph panel indicating lights: The control panel face shall have a graph panel of the system with lights at appropriate positions to indicate:

    a. Cooling fan on

    b. Lime system operations: bin actuator on; feeder on; pneumatic blower on; rotary valve on; high, low bin level

    c. High baghouse differential pressure

    d. Cleaning system on

    e. Cleaning system control power on

    f. Low compressed air pressure

    g. Bypass damper open/closed

    h. Inlet damper open/closed

    i. Outlet damper open/closed

    j. Other indicating lights as recommended by the Seller to properly monitor the baghouse system

    k. Baghouse differential pressure indicator

    l. Audible alarm to annunciate high differential pressure or high inlet temperature

    m. ID fan on

- Control panel face shall include all necessary selector switches and/or push buttons to initiate control and operating functions.
- Alarm conditions including, but not limited to, high inlet temperatures, low inlet temperatures, lime supply failure, high baghouse differential pressure, low compressed air pressure, damper position failure, shall be configured to provide a single dry alarm contact for remote indication.
- Programmable logic controller (PLC): The programmable logic controller shall include three basic components. These

TABLE 8.3. Capital Cost Elements and Factors.*

Cost Elements	ESP	Venturi Scrubbers	Fabric Filters	Thermal and Catalytic Incinerators	Adsorbers	Absorbers	Condensers
Direct Costs							
Purchased equipment cost**	1.00	1.00	1.00	1.00	1.00	1.00	1.00
Other direct costs:							
Foundation and supports	0.04	0.06	0.04	0.08	0.08	0.12	0.08
Erection and handling	0.50	0.40	0.50	0.14	0.14	0.40	0.14
Electrical	0.08	0.01	0.08	0.04	0.04	0.01	0.08
Piping	0.01	0.05	0.01	0.02	0.02	0.30	0.02
Insulation	0.02	0.03	0.07	0.01	0.01	0.01	0.10
Painting	0.02	0.01	0.02	0.01	0.01	0.01	0.01
Total direct cost	1.67	1.56	1.72	1.30	1.30	1.85	1.43
Indirect Costs							
Engineering and supervision	0.20	0.10	0.10	0.10	0.10	0.10	0.10
Construction and field							
expenses	0.20	0.10	0.20	0.05	0.05	0.10	0.05
Construction fee	0.10	0.10	0.10	0.10	0.10	0.10	0.10
Start up	0.01	0.01	0.01	0.02	0.02	0.01	0.02
Performance test	0.01	0.01	0.01	0.01	0.01	0.01	0.01
Model study	0.02	—	—	—	—	—	—
Total indirect cost	0.54	0.32	0.42	0.28	0.28	0.32	0.28
Contingency†	0.07	0.06	0.07	0.05	0.05	0.07	0.05
Total‡	2.27	1.94	2.21	1.63	1.63	2.24	1.76

*As fractions of total purchased equipment cost. They must be applied to the total purchased equipment cost.

** Total of purchased costs of major equipment and auxiliary equipment and others, which include instrumentation and controls at 10%, taxes and freight at 8% of the equipment purchase cost.

†Contingency costs are estimated to equal 3% of the total direct and indirect costs.

‡For retrofit applications, multiply the total by 1.25.

139

components are the processor, power supply, and input/output (I/O) section.

a. The processor shall be a complete solid-state device designed with sufficient capacity for all required inputs and outputs. The main function of the processor will be to continuously monitor the status of all inputs and direct the status of all outputs. The processor will be designed for a hostile industrial environment and a maximum ambient operating temperature of 100°F. It will operate on 120 volts, ±15%. The memory shall be equipped with five-year life lithium batteries to provide DC power to maintain memory whenever there is an extended power failure or power is turned off. The processor shall be provided with the following controls and indicators: run light, power on light, battery on light, and memory protect key.

b. The power supply shall be mounted inside the front cover of the processor and shall not require any adjustments or maintenance. The power supply shall have sufficient capacity to operate the processor and the required number of inputs and outputs.

c. The input/output section shall consist of the various required types and numbers of I/O modules that will be rack mounted by using heavy duty metal housings designed to contain the I/O modules. Each I/O rack will be directly connected to the controller. The I/O modules will be solidly constructed units that are easily removed or plugged into their housing. Once inserted, electrical contact is automatically made through plated spring connectors. There will be no requirements to shut down the system to replace I/O modules.

d. The PLC shall be factory programmed by the CONTRACTOR to accomplish the functions required for proper operation of the system.

8.6 COST FACTORS

Capital cost estimating factors have been developed for air pollution control equipment and are presented in publication EPA/625/6-86/014, titled "Control Technologies for Hazardous Air Pollutants," September 1986. Table 8.3 summarizes these based on delivered, purchased equipment costs. Note that, in general, total installed cost is roughly twice the purchased cost.

CHAPTER 9

Principles of Operation of Fabric Filters

RICHARD BUNDY

9.1 HISTORY

THE first fabric filter system that was commercially sold for use on an industrial boiler was sold in 1973 and went into commercial operation in 1974 at a chemical plant in West Virginia. The installation consisted of four independent systems. It was a perfect example of the design of air pollution control equipment when it is first used on a new application.

The design made use of the supplier's standard baghouse and was modified somewhat in consideration of the demands of the application. The "dust collector" was improved only marginally by using #10 gauge steel instead of #12 gauge, an inch of insulation was installed and it was divided into compartments by internal walls. Although the system works and continues to run today, it is not without problems. A baghouse must be designed for a specific application in order to operate optimally.

Boilers present altogether different operating conditions and requirements than a typical dust collector. In order to design, select, or evaluate a baghouse for this application, one first needs to know something about baghouses in general.

As experience was gained, the design soon evolved essentially to industry standards of 3/16″ plate, 3″ thick insulation, independent modules and a host of other design features.

9.2 THEORY OF OPERATION

9.2.1 COLLECTION MECHANISMS

Filtration of particulate in a fabric filter utilizes four (4) basic mechanisms: direct impaction, interception, diffusion, and electrostatic force. These forces all serve to make a particle hit and stick to a fiber in the fabric.

Particulate collection by impaction occurs with particles large enough to maintain inertia. As the flue gas stream lines curve to pass the individual fibers, the particle has enough inertia to break away from the air stream and travel straight to impact on the fiber where it is trapped.

In interception, a particle follows the air stream line as it curves around the fiber, but it comes close enough to a fiber that it is intercepted as it passes and is caught on the fiber and collected.

Diffusion mechanism affects the smaller particles in the 0.1- to 0.3-micron range. These particles do not follow streamlines, instead they are influenced by Brownian Movement to cross streamlines. As they randomly move while the streamline curves around the fiber, they may impact on the fiber.

Electrostatic attraction between a particle and the fiber or the repelling force between particles can both cause a particle to move to the fiber and adhere.

These four (4) mechanisms all add up to filtration. But filtration by a clean fabric is not highly efficient. There is one additional condition that enhances the filtration process: filter cake.

9.2.2 FILTER CAKE

Although the fabric may look like an impenetrable mat, it is not. In microscopic terms, there are huge gaps between the fibers, which can allow between 2% and 10% of the particulate to escape filtration. As the fabric collects some particulate, it builds what we call a "filter cake." This layer of material fills the voids, and the particulate becomes the filter medium for additional particulate collection, and the collection efficiency increases as the cake builds.

9.2.3 BAG CLEANING

As the filter cake builds, it increases the resistance to flue gas flow, and the pressure drop across the baghouse increases. When the pressure drop reaches the design operating point (typically 2–4″ w.g.), the bags must be cleaned. There are several different cleaning mechanisms employed to remove the particulate. The methodology depends first on the type of baghouse.

9.3 COMPONENTS OF A BAGHOUSE

A baghouse is comprised of a dirty air plenum, a clean air plenum, a hopper, and the cleaning mechanism.

Flue gas flow enters the dirty air plenum (which includes the hopper) and is directed to the filtering medium (bags). The bags and their supporting framework (tubesheet or cell plate) forms the interface between the dirty and clean air plenum. The clean air plenum serves as a manifold, collecting the gas flow from the individual bags and carrying it to the outlet duct.

The hopper is located under the dirty air plenum and serves as the collection point for the particulate. The cleaning mechanism, as discussed below, causes the collected particulate to be removed from the bags on a periodic basis.

In addition to these major components, there are a lot of subsystems such as duct manifolds, isolation and by-pass dampers, ash conveying equipment and a control panel.

9.4 TYPES OF FABRIC FILTERS

There are two (2) major categories of fabric filters: inside bag collectors and outside bag collectors.

An inside bag collector collects the particulate on the inside of the bag, which is attached to a tube sheet (cell plate) at the bottom of the bag. All of the gas flow enters the bag, moves upward into the bag, and flows out thru the fabric. The clean air plenum runs from the bottom of the bag, on the outside, to the top of the housing. The bag is inflated by the flue gas and therefore does not require any support other than suspension from the top.

An outside bag collector captures the particulate on the outside surface of the bag. The bag is suspended from a tubesheet near the top of the housing, and there is no connection at the bottom. Since the bag is sucked in by the pressure, a rigid cage is installed inside the bag to keep it from collapsing. The clean gas exits the top of the bag, and the clean air plenum is only the top of the housing.

The different configurations of inside and outside baghouses dictate two (2) resultant design and operating differences: bag cleaning method, and bag maintenance.

9.4.1 SHAKER/REVERSE AIR VERSUS PULSE CLEANING

Bags in an inside bag collector are cleaned either by a reverse air (back flow) or by shaking. In either case, the bags to be cleaned have to be isolated from the flue gas flow. The system is arranged in multiple modules (compartments), each of which can be isolated by damper valves. This isolation is necessary because any flow through the bag would prevent the particulate from being forced off the bag.

Reverse air cleaning consists very simply of blowing cleaned flue gas into the clean air side of the module and allowing it to flow backwards through the bag. The bags have rings sewn in them every few feet so that they do not collapse; however, they do fold inward. This folding fractures the filter cake, and the backward flow carries it off the bag and into the hopper.

There is one inherent problem with this technique. The reverse gas flows out the module inlet and enters the raw flue gas stream where it then enters the other compartments. The finest of the particulate that was cleaned off the bags does not fall out in the hopper, but instead is recirculated back to another module. In some applications, this recirculation of fines builds up over time and causes a gradual increase in the operating pressure drop, or what is called "creeping pressure drop."

Shaker cleaning of the bags uses the same module isolation technique. The bags are attached at the top to a frame that is rocked back and forth, causing a rippling of the bags that moves down the length, "shaking" the particulate off the bags.

In some applications, a combination of both reverse air and shaking is used to clean an inside bag collector.

Outside bag collectors are cleaned by a variety of techniques, most all of which can be categorized as some type of "pulse." A jet of high pressure air is injected inside the bag and a pressure bubble travels the length of the bag. This bubble causes the bag to rapidly expand to its full diameter at that point. Then the inertial shock as the bag hits its expanded limit causes the collected ash to separate and fall off.

Pulse cleaning is categorized by the pulse pressure and the pulse air volume that is injected for cleaning. They are named low pressure (<5 psig)/high volume cleaning, intermediate pressure ($10–50$ psig)/intermediate volume cleaning, and high pressure (>80 psig)/low volume cleaning. Most baghouse vendors seem to think that their way is the best; however, all have worked well. The different arrangements use different Venturi arrangements, and it is likely that the amount of induced air that is injected and the change in pressure during the pulse makes them all essentially the same. The bag doesn't know or care which system pulsed it.

The bag cleaning can be done with the module either in the filtering mode (on-line) or while isolated (off-line). Even when the module is cleaned on-line, the individual bags being cleaned are effectively taken "off-line" for the brief instant that the pulse enters the bag because it blocks the throat of the bag.

Off-line cleaning is usually more efficient because the complete disruption of gas flow through the module greatly reduces the amount of ash re-entrained; however, either type can work.

9.5 SIDE STREAM SYSTEMS

A distinct fabric filter system used on certain industrial boiler applications is the side stream system or the hopper evacuation system. It is typically used as a retrofit or in cases where a high removal efficiency is not required. The system operates by exhausting a fraction of the flue gas (usually 12–20%) from the side of a mechanical collector and taking it to a baghouse. With the proper take-off design, a negative pressure is created at the mechanical collector tube so that as the spinning flue gas starts to turn up the center of the tube, some of the particulate breaks away from the air stream and is drawn off with the evacuated gas stream.

By taking just 20% of the flue gas, as much as 50% of the particulate can be treated. Typically, systems have operated with outlet emissions of 0.03 to 0.05 lbs/MMBtus. This does not make an efficient enough system for New Source Performance Standards, except on very small units, but it can save a borderline existing unit from a more expensive alternative.

9.6 CRITICAL DESIGN ISSUES

Baghouses of a specific generic type from different suppliers may seem to be very similar. In fact, they are, in many respects. However, there are a few critical areas that make the difference between success and failure of a baghouse system. These issues are in addition to the obvious differences in the housing material thickness, quality of fabrication, selection of proper component suppliers, etc.

The five most significant areas of baghouse design are

(1) Fabric selection
(2) Gas velocity and distribution (air/cloth ratio)
(3) Bag/bag support design
(4) Bag cleaning system
(5) Maintenance considerations

Each of these items is critical to the successful performance and/or the operational and maintenance costs of the system.

9.6.1 FABRIC SELECTION

The bag fabric is regarded as the heart of the baghouse. It is the cause of meeting or failing to meet the environmental performance criteria, it can be the single biggest operational problem if it is not suitable for the application, and its replacement probably will represent the greatest operating cost. The different alternatives that may be selected for boiler applications are shown in Table 9.1.

Selection of the correct fabric initially is a process of elimination.

(1) The boiler provides a certain flue gas temperature. The bag must be capable of operating at that temperature, and any fabrics that cannot are not worth further consideration. Of those shown, usually polyester and acrylic will be eliminated on this basis.

(2) The flue gas chemistry is the second screening criteria. The sulfur content of the fuel and whether or not there is a flue gas scrubbing system will determine the suitability of some fabrics. Nomex would not be acceptable for untreated flue gas from almost any coal; however, when a circulating fluid bed boiler scrubs the SO_2. Nomex has been very successful.

(3) Required collection efficiency is a consideration. Although other factors, such as air-to-cloth ratio, can compensate for the inherent filtration capability of a fabric, it is still common to consider the more efficient fabrics more favorably when a very low outlet emission must be met.

(4) Cost is obviously a consideration. There is a substantial spread in the relative costs of the fabrics. These initial costs have to be considered with the size of the system, the relative merits of the fabric in regard to the above issues, the expected service life, operating hours, reliability, upset potential, etc.

(5) Performance history—most all of these fabrics have been around for some time and, therefore, must have enjoyed some measure of success. However, there are some whose performance has always been marginal, while others generally have met with success.

Besides selection of the base fabric, there are alternative finishes that can be applied. Some of these are chemical coatings (i.e., Teflon coating on fiberglass), while others are mechanical (i.e., singeing a felted fabric).

9.6.2 GAS VELOCITY AND DISTRIBUTION

Air-to-cloth ratio is the primary sizing factor for baghouses. It is determined by dividing the flue gas volume (acfm) by the effective cloth area of the baghouse. It can be stated in both the "gross" or "net" condition.

- gross air-to-cloth ratio = acfm/total cloth area
- net air-to-cloth ratio = acfm/cloth area with a module off-line for cleaning or maintenance

TABLE 9.1. Fabric Alternatives.

Fabric	Temp. Limit	Filtration Efficiency	Abrasion Resistance	Ash Release	Acid Resistance	Relative Price**
Acrylic	284°F	Very Good	Excellent	Good	Good	$15
Fiberglass (felt)	500°F	Very Good	Poor	Good	Fair	$75
Fiberglass (woven)	500°F	Fair	Poor	Fair	Good	$25
Goretex	500°F	Excellent	Poor	Excellent	Good	$100
Nomex	400°F	Very Good	Excellent	Good	Poor	$30
P-84	500°F	Excellent	Fair	Good	Good	$50
Polyester	275°F	Very Good	Excellent	Good	Fair	$10
Ryton	350°F	Good	Excellent	Good	Good	$45
Teflon	500°F	Good	Good	Fair	Excellent	$200

*There are different fabric weights, finishes, and construction techniques that will alter the data for some fabrics.
**Typical price for a 6″ diameter by 10′ bag. Exact prices would vary depending on factors in above note and quantity, fabrication specifications and vendor.

If there is only one compartment, then there effectively would not be a "net" condition.

Besides air/cloth ratio, there are two other sizing issues relative to the gas velocity and distribution.

Can velocity is a relatively new term and one that probably deserves more consideration than it has traditionally received. Can velocity quantifies the internal gas velocity that affects both abrasion (from the dirty gas being filtered) and particle re-entrainment (i.e., returning particles that have been pulse cleaned from the bags back to the bag surface). Can velocity is difficult to define exactly. It is regarded as the maximum velocity of the flue gas where it would wear the fabric or entrain particulate. It is easier to provide examples.

In an inside bag collector, it would occur at the point where the flue gas enters the bottom of the bag. One would calculate it by multiplying the cloth area of the bag by the air-to-cloth ratio (= acfm per bag) and dividing it by the area of the opening in the bottom of the bag. This yields the velocity of the flue gas as it enters the bag.

In an outside bag collector, with a hopper inlet, or any inlet that forces the gas flow into the hopper, the can velocity occurs at a plane even with the bottom of the bags. It is calculated by dividing the gas volume for one module by the cross-sectional area of the module, less the area occupied by the bags. The maximum can velocity that is typically successful on boiler applications is 225 to 250 fpm.

Proper gas distribution is often not given the consideration that it deserves. From the above descriptions, it is obvious that both air-to-cloth ratio and can velocities are calculated as averages. In fact, they are important as absolute values. If gas were unevenly distributed throughout a system that was designed for 4.0:1 air-to-cloth ratio, such that one module operated at 1.0:1 while another operated at 7.0:1, it would not be successful.

Similarly, an average value can velocity of 225 fpm can be designed, but with poor inlet gas distribution, some areas of the baghouse may have a velocity of 500 fpm or more, while others are dead air space or experience reverse flow. This causes very premature bag failure.

9.6.3 BAG/BAG SUPPORT DESIGN

In order to maximize the bag life, it is very important to properly design and install the bag with its supporting structure. In most designs, there is some combined use of hardware and tension to accomplish this.

Inside bag collectors with reverse air cleaning are placed under tension in the baghouse, and rings are sewn into the bags every few feet to keep them from folding inward on themselves. Shaker cleaned bags generally do not have rings, but rely solely on tension. Outside bag collectors almost all use a wire frame or cage inside the bag.

It is beyond the scope of this chapter to discuss the critical points of each possible design, but some overview points are appropriate.

Bag tension generally has a relatively narrow band in which to work. There must be enough tension to prevent the fabric from folding because that would lose filtering area and the folds become weakened and are points of early failure. It also must be great enough to prevent the bag from collapsing during cleaning in a reverse air baghouse. On the other hand, too much tension can put undue stress on the fabric and cause early failure. The potential problem is compounded by the thermal expansion that baghouses on boilers experience. The tension put on a bag during ambient temperature is not the same as it experiences when the baghouse is heated up to 400°F.

Bag support, whether by a cage or sewn-in rings, is done to hold the bag against collapse. It is important to have enough support to meet the mechanical requirements of the fabric versus the operating conditions. For this purpose, the more support, the better. On the other hand, more support costs more money, each supporting member blocks off some amount of fabric from effective filtration, and support points can be sources of wear. Proven performance is the only way to demonstrate that a good balance has been achieved.

9.6.4 BAG CLEANING SYSTEM

The filter media (bags) have to be cleaned on a regular basis to prevent an excessive resistance to flue gas flow (pressure drop), which would reduce draft at the boiler. The bag cleaning has to be efficient enough that it stays even with the incoming ash load, otherwise the pressure would gradually increase. On the other hand, overcleaning a bag can be detrimental for two reasons. The most obvious is that each cleaning of a bag flexes and stresses it, and the more frequently it is done, the shorter the bag's life will be.

The second reason is that overcleaning destroys the filter cake. As described early in this chapter, a clean bag is not a particularly efficient filter. It relies on the build-up of a layer of ash (filter cake) on the bag to help close off the pores of the fabric. Each time a bag is cleaned, there is some destruction of that filter cake. To a reasonable extent, this is normal and something that is considered in the predicted efficiency of the baghouse. Whenever the cleaning frequency exceeds the normal level or its intensity is too great, the filter cake is so damaged that the efficiency is less than acceptable.

There are different cleaning techniques for different baghouses. Under "types of baghouses," we noted the differences between inside bag collectors and outside bag collectors. Regardless of the cleaning technique, there are a couple of basic similarities:

(1) Most boiler or other process gas baghouse applications use off-line cleaning – a module is isolated during the cleaning cycle.

(2) Cleaning is usually initiated by pressure differential. There is a differential pressure set point (usually 2–4″ w.g.) that triggers the cleaning cycle. Most systems have an additional control by an overriding timer that will clean the system regardless of pressure drop, if the pressure set point has not called for it within the normal operating time.

Once cleaning has been initiated, in some cases the program is set to continue cleaning only until enough of the bags have been cleaned to satisfy the pressure set point. In other cases, it is continued until all of the baghouse has been cleaned. The latter case is generally the better; however, there are cases when the first is preferable.

When all of the baghouse is cleaned in one cycle, there is less time that it operates with some compartments having clean bags while others have dirty bags, a situation that causes an imbalance in gas flow.

There are specific cases when partial cleaning is better. If the opacity of the system is only bordering on the acceptable, there may be puffs when the whole baghouse was cleaned that would not be present if the cleaning were spread over several steps. Secondly, when a baghouse is used as part of an acid gas scrubbing system, a significant portion of the acid gas reaction may take place in the filter cake. In this case, total cleaning may result in temporarily high acid gas emissions that would also be smoothed by stepped cleaning.

9.6.5 MAINTAINABILITY

Obviously, a baghouse must be maintained, and because it operates in such a hostile environment, it is critical that the design is such that the maintenance can be done without undue problems. The techniques and issues for inside and outside bag collectors are different.

In an inside bag collector, the bag is attached at the top and the bottom, and the bag length is between fifteen and thirty feet. With the bag attached at both ends, it is necessary to enter the baghouse for bag maintenance. There are internal walkways at both levels, and the bags' area is typically arranged so that there is no bag deeper than three (3) or four (4) rows from a walkway.

There are two disadvantages to this method. First, maintenance is performed inside a very dirty and confining module. Heat, lack of light, dust,

TABLE 9.2. Fabric Filter Cost Summary.

Item	Minimum	Maximum	Typical
	Range		
Baghouse			
Module	$115,000	$125,000	$125,000
Hopper Heaters	0	12,000	12,000
Level det/vibrator	0	6,000	6,000
Insulation	55,000	75,000	65,000
Access/wthr prot	0	47,000	23,500
Bags and cages			
Cages	20,000	75,000	20,000
Bags	28,000	115,000	35,000*
Ductwork and valves			
Ductwork	30,000	30,000	30,000
Manifolds	15,000	15,000	15,000
Inlet valves	11,500	13,000	13,000
Outlet valves	13,000	15,000	15,000
Insulation	15,000	15,000	15,000
By-pass			
Valve	8,000	20,000	8,000*
Duct	2,500	2,500	2,500
System isolation	0	12,000	0*
Air moving system			
Fan and motor	32,000	32,000	32,000
Draft control	3,000	30,000	4,500*
Access			
Top	12,000	20,000	16,000
Pulse valves	0	7,000	7,000
Hopper acc'y	0	10,000	0
Ash removal			
Airlocks	3,500	40,000	14,000*
Conveyor	14,000	30,000	14,000*
Controls			
Base panel	35,000	35,000	35,000
Graphic display	0	2,500	2,500
Recorder	0	2,500	2,500
Total Equipment Cost	$412,500	$786,500	$512,500
Approx. installation cost	192,000	274,000	229,000**
Installed Cost	$604,500	$1,060,500	$751,500**

*Site specific conditions could change the value.
**Typical, for equipment listed—not including foundations, building modifications or engineering.

151

acid fumes and a lack of space to work all make it a very unpleasant task. Second, when only a few bags have failed, the ones that are not in the first row are very difficult to inspect and replace. A bag failure can be any place on the bag, including the back side. The general area of failure is located by looking for signs of contamination (fallout on the floor, markings on the sides of the bag). Then it is necessary to grope and push bags aside to find the exact failure. The bags in front have to be removed to gain access to the bags in the back rows whenever that is the location of the failure.

Outside bag collectors all essentially use some form of top bag access. The top of the baghouse is exposed either thru hinged or removable top doors, or via a tall plenum that personnel can enter (a walk-in clean air plenum). The bags and cages are then serviced from the top, usually in an ambient environment.

Aside from the more pleasant working conditions, there is the advantage that no matter where the leak is in the bag, the leaking ash will exit at the top, making this the only area that has to be inspected.

9.7 SUMMARY

Industrial boilers represent a challenging set of operating conditions for a baghouse air pollution control system. However, the application is becoming a more mature one, and many of the problems of earlier units have led to design changes in today's units that will allow lower maintenance costs and longer lived systems.

It is somewhat of a surprise that these years of experience have not led to more standardization of design. Both of the generic types of baghouses (inside and outside bag collectors) are still being used in addition to a number of variations in bag cleaning design and maintenance considerations. Although there are personal preferences for these items among different suppliers and users, there is no denying that they all work to some extent. With the proper component design and selection, they can each be successful.

Table 9.2 is a summary of fabric filter system costs in 1990 year dollars for industrial size boilers. A range of costs is listed, as well as average costs for typical systems.

CHAPTER 10

The Cost of Control and Retrofits

HOWARD E. HESKETH

10.1 INTRODUCTION

WHEN a system cost is to be determined, one should look at the entire process and consider devices and processes and not just individual components. In other words, we need to do the best job, and this should be done at a reasonable cost. There needs to be funds available to purchase and install the system (capital costs), as well as to operate and maintain the system (annual operating costs). The process should be considered in total. For example, low sulfur coal at $10/ton more than high sulfur coal could be used in a system requiring less scrubbing for SO_2 control. Supply and demand set prices. This could mean low sulfur coal costs will increase more than that of the high sulfur coal. On the other hand, installation of good quality, reliable control equipment may permit burning of the cheaper coal at an overall lower cost.

Some of the biggest operating costs are those for chemicals and those for energy to operate the equipment. Low energy–high efficiency systems are great and usually can pay a return compared to other systems. A thorough understanding of the system chemistry is critical for how much and what chemicals are required and how much and what products are produced. Costs of some different alkali chemicals in 1992 averaged about:

$$\text{limestone} = \$40/\text{ton}$$
$$\text{lime} = \$51.70/\text{ton}$$
$$\text{NaHCO}_3 \text{ (sodium bicarbonate)} = \$220/\text{ton}$$
$$\text{Mg(OH)}_2 = \$250/\text{ton}$$

Chemical reactivity, type of product (e.g., ground, calcined, purity),

153

TABLE 10.1. Generalized Costs in 1988 Year Dollars.*

Item	Material of Construction	Purchase Cost, $	M&S Index	Conditions
Fans, complete	Carbon steel	$42.3d^{1.2}$	830	wheel dia = d; $12.25'' \le d \le 36.5''$; static press = $0.5\text{-}8''$ H_2O
Ductwork	FRP	$53.7d^{1.38}$	830	$10.5'' \le d \le 73''$
	PVC	ad^b	855	dia = d = 6-12''; 14-24''; $a = 0.877;\ 0.0745$; $b = 1.050;\ 1.9800$
Dampers	Carbon steel	$4.84d^{1.50}$	830	$13'' \le d \le 40''$
Cyclones	Carbon steel	$6520A^{0.903}$	855	cyclone inlet area = A, ft²; $0.2 \le d \le 2.64$
Air Lock		$2730A^{.0965}$	855	incl. fan and stand; "

*Source: Vatavuk, W. M., "Pricing Equipment for Air Pollution Control," *Chemical Engineering*, Vol. 97, No. 5, May 1990, pp. 126–130.

quantity used and location relative to source of the chemicals are some of
the factors that affect costs.

10.2 GENERALIZED EQUIPMENT COSTS

This chapter discusses control costs in a general manner. For example,
typical total capital cost of a "large" conventional fuel PC (pulverized coal)
boiler with generator, ducts, FGD control by lime/limestone scrubbing
only and stack is about $750/kW. The FGD system alone could be
$150–200/kW in 1990 year dollars. However, it is likely that costs will
need to be estimated component by component for specific installations.
Costs of emission analyzers may not be included in these dollar amounts.
Estimates for these are provided in Chapter 12. Generalized costs are
listed in Table 10.1.

Table 10.1 is a listing of costs for various components in 1988 year dollars. Fan costs depend on the material of construction and on size (the fan
wheel diameter, d). FRP fans are considerably less costly, but have limitations (e.g., temperature). One can calculate specific charges for the
various components using procedures such as this.

Note that there is a time base index to when these equations apply. The
M&S Index is the Marshall & Swift Index. There are index values for each
quarter of the year. This is to be consistent as the equations are really
relative to a particular time period of each year. Frequently, the Marshall
& Swift Index value used will be an average yearly value. The values of the
cost index historically rise with time. The index references used here
mainly come from publications in the Chemical Industry, mostly *Chemical Engineering*. The *Journal of the Air Waste Management Association*
(AWMA) also reports some of these values. It should be mentioned that
William M. Vatavuk of the USEPA has put together a book (1990) published by Lewis Publishing titled *Estimating Costs of Air Pollution Contol.*

Figures 10.1–10.5 are equipment cost curves. Figures 10.2–10.5 are as of
December 1977, when the Marshall & Swift Index value was 505. Figure
10.1 is for various types of pumps, with different capacity. These costs, in
thousands of dollars, are as of November 1988.

Figure 10.2 shows stack costs. Stacks alone may or may not be charged
as control equipment costs. Stacks may be lined or unlined. Obviously,
lined stacks cost more than unlined stacks. Figure 10.2 gives just two fami-
lies of curves—one for 15-foot and one for 30-foot diameter stacks. The
costs are given as a function of the height. Wet scrubbing systems often require insulated lined stacks, although many systems do not have insulated
stacks. Some systems have a ceramic stack and with proper care, they have
been working well. Stacks without complete control systems are more

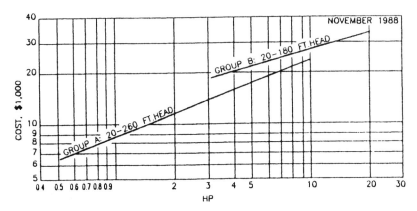

Figure 10.1 Centrifugal pumps—single-stage; 20- to 260-ft head, 316 stainless steel (includes motor and starter) M&S = 866.

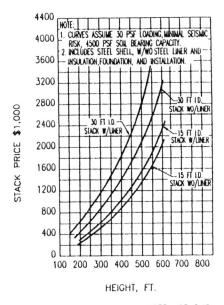

HEIGHT, FT.

PRICES FOR TALL STEEL STACKS, INSULATED AND LINED. M+S=505

Figure 10.2 Tall stack cost estimates.

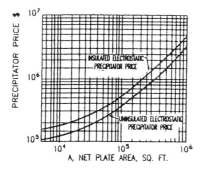

Figure 10.3 Dry type electrostatic precipitator purchase prices vs. plate area, M&S = 505. Note: for high efficiency, ESP SCA, use 400 ft² plate area per 1,000 cfm gas.

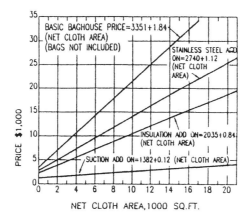

NET CLOTH AREA,1000 SQ.FT.

Figure 10.4 Intermittent pressure, mechanical shaker baghouse prices vs. net cloth area. M&S = 505. Note: typical filter air/cloth ratio is 2.2 ft²/cfm.

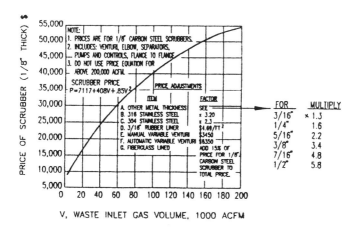

V, WASTE INLET GAS VOLUME, 1000 ACFM

Figure 10.5 1/8″ Thick carbon steel fabricated scrubber price vs. volume, M&S = 505.

157

likely to fail. All stacks need periodic checking and maintenance. With any stack, one needs to check as to whether the cost includes the cost of the foundation base. Imagine what can happen to a high stack in strong winds or during an earthquake.

Figures 10.3, 10.4 and 10.5 indicate cost curves for electrostatic precipitators (ESP), baghouses and scrubbers. ESP's are generalized on the base of specific collecting area (SCA). For efficient ESP operation, use an SCA of about 400 sq ft of surface plate surface area for 1,000 acfm of gas. This is just an average and is the base of Figure 10.3. Knowing the ESP size, i.e., plate sq ft area of the precipitator, Figure 10.3 can be used to obtain the cost. Cost would depend upon whether the ESP is insulated or uninsulated thermally. The Marshall & Swift Index as of 1977 was relatively low (505) as noted earlier.

The baghouse, filter, cost curve is shown in Figure 10.4. Air-to-cloth ratio (A/C) and net cloth area, as has been discussed in previous chapters, are the basic sizing factors. As a hint, if you don't know what to use, start out using an A/C ratio of 2.2. Figure 10.4 is for a shelter type baghouse and gives cost versus size trends. Different material of construction and insulation could add to these costs. Note that costs of filter bags are not included. Bag costs are not capital costs; they are operating costs. These bag costs are encumbered over the life of the bags using the capital recovery factor (CRF). A single pulse jet filter bag, 6″ diameter × 14′ long, costs from $12 to 125, depending on material.

Figure 10.5 shows costs of Venturi scrubbers. There are many different types of scrubbers. Figure 10.5 is for a Venturi elbow separator. Pumps and controls are all included in these costs. The curve shown is for a thin, 1/8-inch carbon steel unit, and it is limited to systems up to 200,000 acfm. The costs are given as a function of inlet gas volumetric flow rate. If one requires thicker material of construction, e.g., 3/16 inch instead of 1/8, multiply by the factor given, i.e., by 1.3 for this example. Going all the way to 1/2″, instead of 1/8″, multiply by 5.8 to adjust your costs. If one decided to make the scrubber out of stainless steel, depending upon which type of stainless steel, use the appropriate multiplication factors given. If you wanted to line it with rubber or make it a variable Venturi throat, again use the appropriate multiplication factors.

10.3 COST EXTRAPOLATION

If you know the purchase cost of a system "similar" to the one that you want to construct, you can go through an extrapolation procedure. This procedure is based on the purchase cost of the known system. The first step is to adjust for size. This is done using the sizing exponent, n. Take the ratio of the desired new system capacity to the original known system capacity to the exponent, *n,* and multiply the result by the purchase cost of the known similar system. This is shown in the following equation. The result is an estimated purchase cost for your new facility if it was purchased at the same time as the old one and made from the same construction materials. The size adjustment extrapolation equation is

Purchase cost of new

$$= \text{(Purchase cost of original)} \left(\frac{\text{new size capacity}}{\text{original size capacity}} \right)^n$$

The sizing exponents, *n,* are obtained historically by plotting data of costs versus size. For common types of control equipment, these values of *n* are approximately equal to

- 0.91 for cyclones
- 0.57 for scrubbers
- 0.63 for ESPs (small sizes)
- 0.72 for filters

The next step is to adjust for material of construction. If the costs are based on a system of 304 stainless steel (S/S), and you want to use 316 S/S or Hastelloy in the new system, check the listing below and note that multiplying by 5.2/4.0 or 7.0/4.0 corrects for material type, but not for material thickness. This can be done using the notes on Figure 10.5.

Adjust for material of construction using ratios of these numbers

- carbon steel   1.0 (base)
- PVC coated steel   1.4
- aluminum   1.5
- lead   1.6
- 304 S/S   4.0
- 316 S/S   5.2
- 317 S/S   6.8
- Hastelloy   7.0

The final adjustment to make would be for time. If the unit you're comparing against was purchased in 1985, you would then take a cost index value from that year and divide that into the current value of the cost index. This is the cost escalation multiplier for time.

In equation form, the adjustment for time is:

$$\text{Purchase cost in yr 1} = \text{(Purchase cost in yr 3)} \left(\frac{\text{M\&S of yr 1}}{\text{M\&S of yr 2}} \right)$$

where M&S is the Marshall and Swift Process Industries average, as printed in the journal *Chemical Engineering* and summarized here:

Year	Mid Year Average M&S Value
1983	761
1984	780
1985	790
1986	798
1987	814
1988	852
1989	895
1990	915
1991	931
1992	943
1993	964

Other cost indices are available. Cost index values are fixed by supply and demand requirements. If you have more people selling than buying, the costs may not escalate, according to a projected cost index value. Also, if you have improved the control device so that it performs better, or if the supplier is able to produce the control unit at a lower cost, this would introduce other factors to be accounted for.

10.4 ESTIMATING CAPITAL COST FROM PURCHASE COST

Purchase cost plus installation cost is capital cost. So far, we have been dealing with equipment purchase cost. It is possible to approximate capital costs using the equipment purchase costs. One procedure is to take the purchase cost and multiply it by an installation factor, which includes costs of delivery, taxes and all other installation charges.

Historically, in the chemical industry and in the utility industry, about half of the capital costs are expended in getting the piece of equipment to your door, and it requires another half to get the unit ready to run (foundations, supports, wiring, piping). When everything is ready, you should be able to push the button and have it run.

For estimating purposes, to obtain capital cost, simply multiply purchase cost by the following factors:

Purchase Cost of	Multiply by
Scrubbers	1.91 = Total Capital Cost
ESPs	2.24 = Total Capital Cost
Filters	2.17 = Total Capital Cost

Assume you have a 20,000 acfm carbon steel wet scrubber, purchased in 1986 for $100,000. What would be the installed cost in 1992 year dollars of a 15,000 acfm 316 stainless steel scrubber of similar configuration?

Solution: Make your adjustments item-by-item. Use the original cost for the 20,000 acfm size system. Your system isn't quite as big, it is 3/4 of the original system size; however, costs are going to be 3/4 to the exponent 0.57. Then, you have material of construction to account for. The initial one was carbon steel and the new one is to be 316 S/S. So costs will be 6.8 times more. Then factor in the cost index value for 1986, and the cost index value for 1992. This is summarized in the following equation where we will adjust original purchase cost for size, material of construction, and time; then account for installation costs:

$$\text{Installed cost in } 1992 = (100,000)\left(\frac{15,000}{20,000}\right)^{0.57}\left(\frac{6.8}{1.0}\right)\left(\frac{943}{798}\right)$$

$$= \$1,300,000$$

With the installation factor of 1.91 included, the estimate shows the answer to be \$1.3 million for the installed cost for a 15,000 cfm unit, out of 316 stainless steel in 1992.

10.5 ANNUAL OPERATING COSTS

One of the most costly operating factors is likely to be the direct cost of chemicals – how much lime, limestone, sodium hydroxide, whatever it happens to be that you require and have to purchase. Run a mass balance, including a stoichiometric factor (i.e., how much chemical above stoichiometric demands will be needed). Operating labor is another direct cost factor. Previous sections of this book discuss hours for the operator, operating labor rates, and how many hours per year you are going to operate. This second direct cost factor, operating labor, may not be high when compared to chemical and power costs. Operating supervision for pollution control equipment is usually about 15% of the operating labor. So, whatever you determine your operating labor to be, add another 15% for supervision.

Utility costs are the other big direct cost with electricity being the largest (unless the system works with steam ejectors or some other driving medium instead of fans and blowers). Power cost is mainly a function of pressure drop. With volumetric flow rate, the kilowatt hours needed can be estimated. The cost, at about 6 cents a kilowatt hour, and the operating hour per year gives the annual cost for electricity. In a like manner, you can calculate the cost of pumping the liquid if it is a wet control system. Relatively speaking, this will be a small cost (e.g., 10% or less of the gas phase pumping cost). Steam may or may not be required. If you use no steam, these costs will be \$0. If it is a wet system, you must pay for water. You may have a sewage disposal charge attached to the water and that would be part of the operating cost.

Maintenance labor is a direct operating cost. The hourly rate for maintenance labor is usually higher than the hourly rate for operating labor. This value times hours per year times the number of maintenance personnel yields the maintenance operating cost. The maintenance material costs for control systems are historically essentially equal to maintenance labor changes.

The other direct cost that we need to look at is the replacement cost for expendable items. In pollution control equipment, certain items (e.g., bags in the filter system) are expendable. When you calculate capital cost of a filter, the bags are not included. Bags will last typically three years, and so you run a capital recovery factor calculation at the company required rate of return (e.g., 10%), and calculate a value that can be dispersed each year for the next three years. This would cycle again for the following three years, so it goes on indefinitely or until the life of the system is reached. As another example, scrubber packing is not part of capital costs. Packing would have a capital recovery factor that could be spread out over about a five-year life. Costs of three expendable items are replacement costs, which are direct annual costs.

Indirect annual costs include capital recovery of the capital expenditure. Recovery of capital (i.e., the installed cost) is the largest indirect operating cost. All equipment must have a life that is acceptable to IRS for tax purposes. Ten to twenty years is a typical lifetime for baghouses, scrubbers, precipitators, and mechanical devices. Some will last longer. Use the capital recovery factor at the required rate of return for the number of years of life that the device will provide. This is the capital recovery portion of annual costs. Administration is usually about 2% per year of capital costs, as a rough rule of thumb. Overhead and "taxes and insurance" are each estimated to be about 2% of capital costs. The total of these direct and indirect annual costs is the annual cost for the system. These can be further broken down to provide cost per kilowatt. Cost of waste disposal may need to be included, and this could be a large operating cost.

10.6 REFERENCES

1 Vatavuk, W. M. 1990. "Pricing Equipment for Air Pollution Control," *Chemical Engineering,* 97(5):126–130.
2 Hall, R. S. et al. 1988. "Estimating Process Equipment Costs," *Chemical Engineering,* 95(17):66–75.
3 Neveril, R. B. et al. 1978. "Capital and Operating Costs of Selected Air Pollution Control Systems—II," *Journal of the Air Pollution Control Association,* 38(8):963.
4 Ibid. 1978, 38(8):829.

CHAPTER 11

Disposal of Residuals from Industrial Boilers

PATRICK WALSH

A LL industrial boiler systems generate residues that must be properly managed to protect health and the environment. All systems produce fly ash and bottom ash, and some systems also produce wastewater. This chapter will discuss alternative methods for addressing these residue control problems.

11.1 TYPES OF RESIDUALS

Industrial boilers and their emission control systems produce a variety of residues. By far, the largest quantity is bottom ash, the unburned and nonburnable materials that drop from the bottom of the combustor at the end of the burning cycle.

The process also produces a lighter emission known as fly ash. Fly ash consists of gaseous components and particulates that are produced either as a result of the chemical decomposition of burnable materials or are non-burned (or partially burned) materials drawn upward by thermal air currents in the incinerator.

Fly ash normally comprises only a small proportion of the total volume of residue from an industrial boiler: quantity ranges from less than 1% to more than 15% of the total ash. Distribution of bottom and fly ash is largely influenced by the type of combustion unit. Excess air systems produce the most fly ash; controlled air units produce the smallest amounts.

Figure 11.1 shows ash moving through a baghouse equipped incinerator. This mass-flow diagram assumes the ash is handled dry. Many facilities water-cool the ash for safe handling, which may result in doubling the weight of the ash to be disposed.

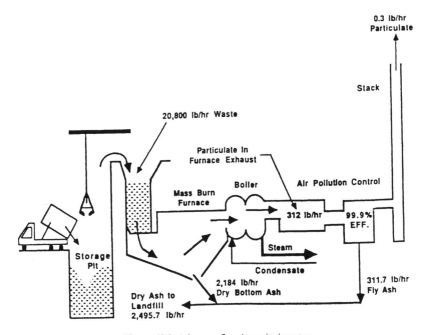

Figure 11.1 Ash mass flow in an incinerator.

Emissions control devices such as electrostatic precipitators and baghouses do not rely upon added materials to control pollutants. Some, however – including dry scrubbers – inject absorbing materials to capture and neutralize acid gases and other pollutants. Such a system could use a fine spray of either calcium- or sodium-based alkaline slurry.

In such systems, the resultant pollutant/slurry mixture is known as scrubber product. Scrubber product must also be properly handled in order to avoid environmental pollution.

11.2 LEGAL CLASSIFICATION OF RESIDUALS

Constituents in both ash and scrubber product vary, depending upon the materials burned. In systems burning a homogeneous fuel, such as coal, oil or tires, levels of pollutants in residuals may be relatively constant. Systems burning a more heterogeneous mixture, such as municipal, industrial or medical waste, may experience wide swings in the chemical composition of residuals.

Combustion efficiency (or burn-out rate) will also affect chemical characteristics of ash and scrubber product. The higher the burn-out rate, the lower the percentage of unburned organics remaining in the residue.

Because significant quantities of hazardous materials may be present in incinerator ash and scrubber product, a test burn in a facility similar in design to the proposed one may be required as a prerequisite to a license. Ash characterization may also be required during facility shake-down and at various times during facility life.

11.2.1 WASTE CHARACTERIZATION

Pursuant to U.S. Environmental Protection Agency (EPA) regulations, a waste material can be classified as legally hazardous if the waste type is liated as hazardous in tables promulgated by the EPA or if the waste type exhibits certain defined characteristics. The four characteristics are ignitability, reactivity, corrosivity or EP toxicity (toxicity demonstrated through performing a pollutant extraction procedure).

Boiler ash is not a listed hazardous waste. To determine if the ash is hazardous, the ash must be investigated to determine if it exhibits a hazardous waste characteristic. The facility operator is required to make this determination.

Since ash is generally fairly inert material, the characteristic of major concern is EP toxicity, which determines the tendency of the waste material to leach certain chemical constituents into the groundwater. To determine EP toxicity, the EPA promulgated a standardized test known as the EP test to determine whether a waste would be classified as hazardous.

The EPA has recently replaced the EP test with a new test, known as the Toxicity Characteristic Leaching Procedure (TCLP). In the future, the TCLP will be utilized to determine whether ash and scrubber product will be classified as either hazardous or solid waste. The new TCLP rules expand the list of residue constituents that must be sampled for (see Table 11.1). Since the TCLP rules are being phased in, facility operators should check with local regulatory officials to determine whether to use the EP Toxicity Test or the TCLP.

11.2.2 PERFORMING THE TCLP

EPA testing procedures expose ash to an acidic environment in an attempt to predict which materials in the ash could ultimately be released into the groundwater. For ash, the test mainly determines the leaching characteristics for heavy metal constituents – since in a well-run facility, organics will be consumed in the burning process.

The TCLP will require the following. First, the waste sample must be crushed, ground, or broken into sufficiently small particles to pass a 3/8-inch sieve. The sample will then be mixed with a pH 5 acetate buffer. If the waste is highly alkaline, an additional acetic acid solution must be added.

The waste/acid solution is then mixed in a tumbler for eighteen hours at

TABLE 11.1. Toxicity Characteristic Constituents and Regulatory Levels.

	EPA HW Number*	Constituent	CAS Number**	Regulatory Level (mg/L)
Metals	D004	Arsenic	7440-38-2	5.0
	D005	Barium	7440-39-3	100.0
	D006	Cadmium	7440-43-9	1.0
	D007	Chromium	7440-47-3	5.0
	D008	Lead	7439-92-1	5.0
	D009	Mercury	7439-97-6	0.2
	D010	Selenium	7782-49-2	1.0
	D011	Silver	7440-22-4	5.0
Volatiles	D018	Benzene	71-43-2	0.5
	D019	Carbon tetrachloride	56-23-5	0.5
	D021	Chlorobenzene	108-90-7	100.0
	D022	Chloroform	67-66-3	6.0
	D028	1,2-Dichloroethane	107-06-2	0.5
	D029	1,1-Dichloroethylene	75-35-4	0.7
	D035	Methyl ethyl ketone	78-93-3	200.0
	D039	Tetrachloroethylene	127-18-4	0.7
	D040	Trichloroethylene	79-01-6	0.5
	D043	Vinyl chloride	75-01-4	0.2
Semivolatiles	D023	o-Cresol	95-48-7	200.0†
	D024	m-Cresol	108-39-4	200.0†
	D025	p-Cresol	106-44-5	200.0†
	D026	Cresol		200.0†
	D027	1,4-Dichlorobenzene	106-46-7	7.5
	D030	2,4-Dinitrotoluene	121-14-2	0.13‡
	D032	Hexachlorobenzene	118-74-1	0.13‡
	D033	Hexachloro-1,3-butadiene	87-68-3	0.5
	D034	Hexachloroethane	67-72-1	3.0
	D036	Nitrobenzene	98-95-3	2.0
	D037	Pentachlorophenol	87-86-5	100.0
	D038	Pyridine	110-86-1	5.0‡
	D041	2,4,5-Trichlorophenol	95-95-4	400.0
	D042	2,4,6-Trichlorophenol	88-06-2	2.0
Pesticides	D020	Chlordane	57-74-9	0.03
	D016	2,4-D	94-75-7	10.0
	D012	Endrin	72-20-8	0.02
	D031	Heptachlor (and its hydroxide)	76-44-8	0.008
	D013	Lindane	58-89-9	0.4
	D014	Methoxychlor	72-43-5	10.0
	D015	Toxaphene	8001-35-2	0.5
	D017	2,4,5-TP (Silvex)	93-72-1	1.0

* Hazardous waste number.
** Chemical abstracts service number.
† If o-, m-, and p-cresol concentrations cannot be differentiated, the total cresol (D026) concentration is used.
‡ Quantitation limit is greater than the calculated regulatory level. The quantitation limit therefore becomes the regulatory level.

a temperature of approximately 22°C (about 72°F). The solution is then poured through a glass fiber filter with a filter pore size of 0.6 to 0.8 microns. The solution passing through the filter is then tested to determine heavy metal content. To test for volatile organics, a device known as a zero head space extractor (ZHE) is used to keep the sample from being exposed to air.

Testing waste samples for a wide variety of pollutants at low concentrations will be both technically difficult and costly. The results of the testing will determine whether fly ash, bottom ash and scrubber ash will be considered solid waste or hazardous waste. Especially for facilities that burn heterogeneous waste types, the protocols for developing a representative sample of ash are crucial to obtaining reliable test results. Individual samples could be high in one or more constituents, causing an entire batch of ash to be declared hazardous waste. Protocols for sampling should be carefully developed with regulatory officials.

11.2.3 MIXING FLY ASH AND BOTTOM ASH

For many systems, bottom ash has generally tested nonhazardous. Fly ash, however, which comprises a much smaller volume of the total, more frequently tests as hazardous. This is becoming increasingly true now that the efficiency of emissions control systems has become high.

If bottom ash and fly ash are mixed together, the mixture can be nonhazardous. This situation has created an interesting regulatory problem concerning whether the mixing of fly ash and bottom ash should be allowed and, if so, how the mixing should be characterized from a regulatory point of view.

EPA hazardous waste regulations normally would define the mixing of a hazardous waste with a nonhazardous waste to create a nonhazardous waste as treatment. Hazardous waste treatment requires a separate license.

However, regulatory authorities have also determined that treatment is an activity performed on a waste after the waste leaves the process which creates it. If fly ash and bottom ash are mixed as part of the process, rather than after the end of the process, the need for a hazardous waste treatment license can be avoided.

For most facilities, the determination as to whether mixing of fly ash and bottom ash is considered treatment is a judgment call left to local regulatory officials. The states generally take different views as to how this dilemma should be resolved. Some states prohibit the mixing of fly ash and bottom ash, requiring fly ash to be taken to a hazardous waste facility.

11.3 HANDLING AND TRANSPORTING ASH

Fly and bottom ash contain pollutants, which could cause respiratory problems to employees handling this residue. Special precautions should be taken to minimize employee exposure risk, including wetting the ash to reduce dust and use of face masks when handling residue in a dry form. New federal and state employee right-to-know and hazard communication regulations are making employee safety and training a priority.

Transporting ash to disposal facilities may be regulated. Most often, covered transport vehicles are required. If the ash is hazardous, special regulations apply, including the need to use a hazardous waste manifest and transporters licensed to haul hazardous waste.

11.4 ASH DISPOSAL

Federal and state regulations require ash tested as hazardous to be disposed of as a hazardous waste. This normally means it will go to a secure landfill known as a Subtitle C facility (named for the hazardous waste regulations that are part of RCRA Subtitle C). These facilities have extensive engineered systems to contain pollutants (see Figure 11.2). Not every state has a Subtitle C hazardous waste landfill. For facilities in these states, the need to haul hazardous ash to a distant landfill increases already high ash management costs.

Disposal options for ash which tests as nonhazardous depend upon local regulations. Some states allow ash to be codisposed with municipal refuse in a sanitary landfill. Others require development of a monofill or monocell (see Figure 11.3).

A monofill is designed and constructed solely to take ash from the facility. Disposal of ash in a monofill avoids exposing the ash to the acidic en-

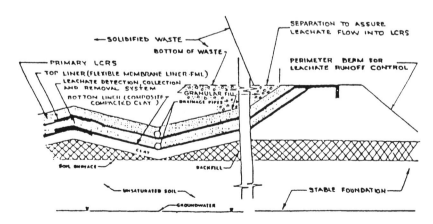

Figure 11.2 Double liner and leachate collection system.

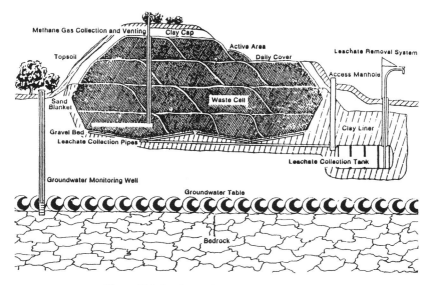

Figure 11.3 Leachate collection engineered landfill.

vironment present in a municipal waste landfill, reducing the potential for leaching pollutants. Also, owning and operating a monofill ensures access to a licensed disposal site for the life of the facility.

As with any state-of-the-art landfill, the monofill should be designed to prevent leachate from having an adverse impact on local groundwater quality. Problems that may exist in municipal landfills, such as the generation of explosive gases, are not present in the monofill due to the ash's inert characteristics.

Some states are requiring the ash be treated to stabilize pollutants prior to disposal. Treatment may involve adding materials to alter the acidity level of the ash, decreasing the potential for leaching. Vitrification, or glassifying the ash, is also being investigated for economic and technical feasibility.

11.5 ASH AS A BUILDING MATERIAL

Ash has also been proposed for use in construction materials. For example, ash has been used to construct concrete blocks and other structural materials. Bottom ash has been used for ice control and fill. Some ash has also been landspread as a soil amendment.

The major question with these reuse technologies is whether pollutants from the ash, especially heavy metals, will eventually leach into groundwater supplies. Testing and analysis is currently under way around the country.

11.6 WASTEWATER MANAGEMENT

While ash is generally the major residue problem at an industrial boiler facility, some plants also generate wastewater. This also must be properly handled in order to protect the environment. Those considering construction of an industrial boiler facility should anticipate and acquire all necessary permits which may be needed for wastewater disposal.

Wastewater from an industrial boiler facility can be generated in various forms. These include system wash water, ash quench water, and water from pollution control systems. These systems also must deal with normal problems experienced by all large industrial facilities, including sanitary wastewater disposal and surface water run-off.

For most small-scale facilities, wastewater can typically be recycled in a closed-loop system. In these systems, water from floor drains, ash dewatering, water softener recharge, and wet emissions control systems are collected and stored in a surge tank. This water is then reused for ash quenching. Sanitary waste can be directed to sewer systems.

A large-scale facility will usually follow the same procedure. However, some facilities allow the water to be absorbed by the bottom ash, which is then taken to a landfill. Other plants place the quenched ash in a holding pond, to allow water to evaporate. With increased regulation of the water pollution potential of ash, holding ponds may be more strictly regulated in the future, perhaps requiring clay or synthetic liners and leachate collection.

For most facilities, the quantity of water used amounts to a few gallons per ton of material burned. Usually this effluent may be discharged to a sewer system. In some cases, regulatory authorities may require that the waste stream be pretreated before discharge.

State regulatory agencies and local sanitation officials should be consulted to determine the best method of handling wastewater.

CHAPTER 12

Combustion and Environmental Emission Monitoring

HOWARD E. HESKETH

12.1 INTRODUCTION

EFFECTIVE, in-compliance operation requires flue gas monitoring in addition to measuring combustion feed rates, ash rates, temperatures and pressures. Figure 12.1 shows an incineration system as an example of a system with detailed monitoring requirements. This figure depicts a rotary kiln incinerator. In the feed, we measure basic chemical composition, ash, Btu content, metals, chloride, and weigh the total material being charged. For the auxiliary fuel, we measure the flow of the fuel. If it is a liquid or gas of known constant composition, this is very easy. It becomes more complex for solid fuels and requires mass and chemical data. It may be necessary to analyze the combustion air to be sure that we do not put any other pollutants in with the combustion air itself. The "F" symbols in the figure are for flow rate. "T" is temperature, and "P" and "ΔP" are pressure values.

Leaving the combustor, measurements are noted for exit flue gas temperature, flow, oxygen, and perhaps CO at this point. Temperature coming out of heat recovery could be critical to operation costs.

This system shows a Venturi, followed by a packed tower for the air pollution control (APC) system. A Venturi here could serve different purposes. It could be for the purpose of quenching the liquid, it could be for the purpose of taking out particles, or it could be expected to do both, depending on the needs and how the unit is operated. Emissions leaving the absorber enter the stack. The water for scrubbing, depending on where it comes from, may require chemical analyses before being introduced into the system. The liquid flow from the wet scrubber may require measuring pHs, POHCs and metals plus leaching tests. Pressure drop across the Ven-

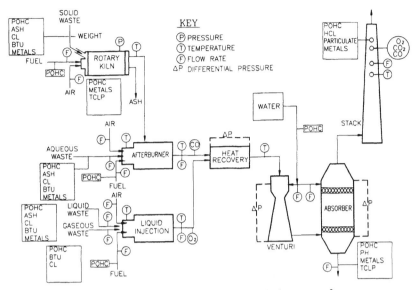

Figure 12.1 Potential sampling points for assessing incinerator performance.

turi plus temperatures, flow rates, oxygen, CO_2, CO, HCl, and particulate metals are all important data and should be recorded on a system such as this.

CO_2 measurements following a wet scrubber do not serve as accurate measurements of CO_2 coming out of a combustion system. Some of the CO_2 can be removed in the scrubbing portion of the system. You may need to make a correction for this. It is better to use CO_2 measurements made after the heat recovery portion of the system (i.e., before the scrubber).

12.2 FAIL SAFETY

Some of our measurements are important enough that they should be used to trigger fail-safe, corrective action. These are noted in Table 12.1. Some of these are for the purpose of protecting the workers, some are for the purpose of protecting from excess emissions, and some are for the purpose of protecting the equipment. High CO in the stack gas may mean that the flame volumetric space or grate area is inadequate, not high enough temperatures, too low a flow of fuel or too high a flow of noncombustible material. Low chamber temperature will normally occur in parallel with high CO emissions. This could be the result of choking the furnace with excess fuel or not providing enough surface or volume heat release area.

The pH of the scrubber water must be checked to protect the equipment and to assure the presence of adequate alkali chemical for reacting with acidic material. Alkalis are required to reduce emissions of acid gases.

Adequate scrubber water flow rate protects from excess emissions and also protects the equipment. Some equipment is constructed of plastic components. As an example, if the flue gases enter a plastic packing or a vessel with an FRP shell and the gases are too hot, the entire APC system will quickly fail. Some installations have three levels of water protection: the recycle liquid loop, the fresh water loop and a fire water backup.

Scrubber pressure drop relates to operating costs. In general, you get what you pay for, i.e., the more pressure drop you put into a scrubber, the better the job is for that particular scrubber for controlling emissions. Sump levels with too little liquid may not be good. We may need minimum amounts of residence times in delay-holdup tanks to enable reactions to go to completion. In addition, do not let these tanks get low, because you want to be able to maintain the scrubber recycle liquid flow rate.

High combustion chamber pressure could be a worker safety problem. High chamber temperature would also be a problem for the equipment. When we talk about combustion chamber temperature, we are talking about temperatures of up to about 2100–2200°F. When we go over 2200°F, this normally indicates a grate temperature or a flame temperature. On the other hand, when we talk about high chamber pressures, we are only talk-

TABLE 12.1. **Parameters Typically Employed to Trigger Fail-Safe Corrective Action.**

Parameter	Basis for Corrective Action		
	Excess Emissions	Worker Safety	Equipment Protection
High CO in stack gas	X	X	X
Low chamber temperature	X		
High combustion gas flow	X		
Low pH of scrubber water	X		X
Low scrubber water flow	X		X
Low scrubber ΔP	X		
Low sump levels	X		X
High chamber pressure		X	
High chamber temperature		X	X
Excessive fan vibration		X	X
Low burner air pressure	X		
Low burner fuel pressure	X		
Burner flame loss	X	X	X

ing about tenths of an inch of water *below* atmospheric pressure for most current systems.

Excessive fan vibration results in equipment failure. In the extreme, fans can jump off their mounts. So this relates to equipment protection, as well as personnel protection. Low burner air pressure, triggers CO (which is a combustible gas) and low chamber temperature which are dangerous. Low burner fuel pressure would cause similar results. If the burner flame goes out, the entire unit must be shut down to prevent emissions and to protect workers and equipment.

12.3 PREDICTING FLUE GAS QUANTITY AND COMPOSITION

This section extends the procedures presented in the combustion chapter, Section 3.7. (Review this section now if necessary.) We took one mole of a dry organic fuel, which happens to represent cellulose. At stoichiometric combustion, we can go ahead and balance our equation. This is shown for 10% excess air and is figured out for 100 pounds of dry fuel. How many moles of flue gas there would be? The answer as shown in the step-by-step outline of Table 12.2 is 20.75.

If we want to predict the composition of the wet gas, we can again take into account water, and we can make a balance showing what total composition we would expect. In this example, the fuel is cellulose and has a very high carbon content. As shown below, we have 15.8% water, very little excess oxygen, and 66.5 mole percent (which for ideal gases is the same as volume percent) of nitrogen. The total of the wet gas adds up to 100%. This is what we get from the total flue gas analysis. If you ran an Orsat type analysis, you would obtain a dry gas analysis.

The calculations for predicting composition of the flue gas (wet) are:

Note: We must adapt the 10% x's air equation to account for the moisture in 1 lb mole dry fuel (162 lb). There would be

$$\frac{9 \text{ lb } H_2O}{90 \text{ lb fuel}} (162) = 16.2 \text{ lb free water or}$$

$$\frac{16.2}{18} = 0.9 \text{ moles extra } H_2O$$

Then corrected wet flue gas analysis would be:

$$\frac{6 \text{ moles } CO_2}{6 + 5 + 24.8 + 0.6 + 0.9 \text{ moles total}} (100) = 16.1\% \ CO_2$$

TABLE 12.2. Step-by-Step Procedure for Predicting Flue Gas Quantity and Composition.

Using a combustion example of 1 mole *dry* organic fuel @ stoichiometric combustion

$$C_6H_{10}O_5 + 6 \left(\frac{79}{21}\right) N_2 \rightarrow 6CO_2 + 5H_2O + 22.6N_2$$

and @ 10% excess air (i.e., 11 × O_2 & N_2)

$$C_6H_{10}O_5 + 6.6O_2 + 24.8N_2 \rightarrow 6CO_2 + 5H_2O + 24.8N_2 + 0.6O_2$$

So per 100 lb dry fuel:

$$(100) \frac{6 + 5 + 24.8 + 0.6}{\dfrac{162 \text{ lb}}{(1 \text{ mole}) \text{ mole } C_6H_{10}O_5}} = 22.5 \text{ moles flue gas per 100 lb dry fuel}$$

Now—if true fuel contains 9.0 lb H_2O + 90 lb cellulose + 1.0 lb ash, then the flue gas per 100 lb as-fired fuel contains:

$$\frac{9 \text{ lb}}{18 \text{ lb/mole}} = 0.50 \text{ mole free water}$$

$$(22.5) \, 0.9 = 20.25 \text{ moles product gases}$$

Flue gas total = 20.75 moles per 100 lb as-fired fuel

$$\frac{5 + 0.9}{37.3} \ 100 \ = \ 15.8\% \ H_2O$$

$$\frac{0.6}{37.3} \ 100 \ = \ 1.6\% \ O_2$$

$$\frac{24.8}{37.3} \ 100 \ = \ 66.5\% \ N_2$$

$$\text{Total} \ = \ 100.0\%$$

Under normal combustion conditions, the flue gases behave as ideal gases, therefore *mole fractions equal volume fractions* and ideal gas laws can be used, e.g., 1 pound mole @ STP = 359 ft³ (Note: Standard Temp. and Pressure (STP) = 32°F and 1 atmosphere). As an example, the quantity of flue gas produced at a fuel feed rate of 1,500 lb/hr (assuming no air in-leaks) at two different temperatures can be estimated by:

@ 2000°F:

$$\left(\frac{1500 \ lb}{hr}\right)\left(\frac{20.75 \ moles}{100 \ lb}\right)\left(\frac{359 \ ft^3}{mole}\right)\left(\frac{460 + 2000}{460 + 32}\right)\left(\frac{hr}{60 \ min}\right) = 9310 \ acfm$$

or

@ 200°F:

$$(1500)\left(\frac{20.75}{100}\right)\left(\frac{359}{60}\right)\left(\frac{460 + 200}{460 + 32}\right) = 2,500 \ acfm$$

The estimate of volumetric flow rate can be made in various parts of the system, assuming no leaks and based on the dry standard cubic feet per minute (dscfm) being constant. Then you can add water at various points, or take out water at various points, or take out the pollutant components as in the real system. Therefore, to simplify calculations, use the dry gas analysis as the base. These calculations are shown as:

(1) Emission rate as dscfm

$$(9310 \ acfm \ @ \ 2000°F)\left(\frac{100 - 15.8 \ moles \ dry \ gas}{100 \ moles \ wet \ gas}\right)$$

$$\times \left(\frac{460 + 68}{420 + 2000}\right) = 6,270 \ dcfm$$

(2) Dry gas analysis: This would be equivalent to an Orsat analysis. For our example:

$$(16.1) \ \frac{100}{100 - 15.8} \ = \ 19.1\% \ CO_2$$

$$(1.6)(1.19) \ = \ 1.9\% \ O_2$$

$$(66.5)(1.19) \ = \ 79.0\% \ N_2$$

$$\Sigma \ = \ 100.0\%$$

12.4 STACK SAMPLING

Assuming sampling is required for particles, as well as velocity and basic combustion gases, the following uses USEPA "Method 5" test procedures (these incorporate "Methods 1, 2, 3") as described in 40 CFR 61. Establish first how many traverse points you have to take using Figure 12.2 for particles. Then divide your stack cross section area into equal area segments and sample each quadrant of the equal area segments, as shown in

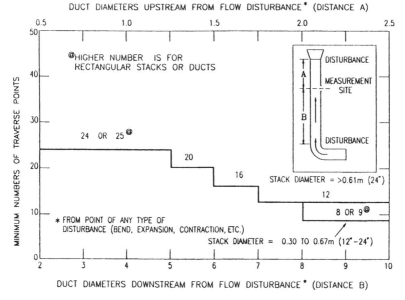

Figure 12.2 Minimum numbers of traverse points for particulate traverses.

TABLE 12.3. Location of Traverse Points in Circular Stacks
(Percent of Stack Diameter from Inside Wall to Traverse Point).

Traverse Point Number on a Diameter	Number of Traverse Points on a Diameter											
	2	4	6	8	10	12	14	16	18	20	22	24
1	14.6	6.7	4.4	3.2	2.6	2.1	1.8	1.6	1.4	1.3	1.1	1.1
2	85.4	25.0	14.6	10.5	8.2	6.7	5.7	4.9	4.4	3.9	3.5	3.2
3		75.0	29.6	19.4	14.6	11.8	9.9	8.5	7.5	6.7	6.0	5.5
4		93.3	70.4	32.3	22.6	17.7	14.6	12.5	10.9	9.7	8.7	7.9
5			85.4	67.7	34.2	25.0	20.1	16.9	14.6	12.9	11.6	10.5
6			95.6	80.6	65.8	35.6	26.9	22.0	18.8	16.5	14.6	13.2
7				89.5	77.4	64.4	36.6	28.3	23.6	20.4	18.0	16.1
8				96.8	85.4	75.0	63.4	37.5	29.6	25.0	21.8	10.4
9					91.8	82.3	73.1	62.5	38.2	30.6	26.2	23.0
10					97.4	88.2	79.9	71.7	61.8	38.8	31.5	27.2
11						93.3	85.4	78.0	70.4	61.2	39.3	32.3
12						97.9	90.1	83.1	76.4	69.4	60.7	39.8
13							94.3	87.5	81.2	75.0	68.5	60.2
14							98.2	91.5	85.4	79.6	73.8	67.7
15								95.1	89.1	83.5	78.2	72.8
16								98.4	92.5	87.1	82.0	77.0
17									95.6	90.3	85.4	80.6
18									98.6	93.3	88.4	83.9
19										96.1	91.3	86.8
20										98.7	94.0	89.5
21											96.5	92.1
22											98.9	94.5
23												96.8
24												98.9

Source: *Federal Register,* Vol. 42, No. 160.

Table 12.3 for circular ducts. Sampling in this nature would be done using probes or pitubes with temperature sensors and nozzles that fit on the end of the probe. The sampling procedure is critical. The nozzle itself can throw out particulate matter that would mean a loss to the sample sizing system if it were not accounted for. This is okay for USEPA "Method 5" type testing, where you wash the probe and the nozzle and combine all the catch, then the dry mass is added to the filter collection mass. This is not the procedure used for continuous emission monitors (CEM), where the sampling tubes are in the stack on a continuous basis. Stainless or glass sample tubes are good for Method 5. However, you do not want to use tubes with glass linings for CEM, because glass absorbs and reacts with SO_2 and gives erroneous readings.

12.5 CONTINUOUS EMISSION MONITORING

There are many instruments that are available commercially to give you readouts. One of the most critical parts of the entire system is getting the gas samples accurately, without contamination, without losses, and conditioned so they do not adversely affect the instruments. The analyzer instruments represent many different devices for different compounds as shown in Table 12.4. Gas chromatography can be used for most compounds, but you have paramagnetic and zirconium oxide electrocatalytic analyzers for O_2. Non-dispersive infrared analyzers are good for CO_2 and CO on a continuous emission monitoring system. Chemiluminescence is common for NO_x. Flame photometry, pulse fluorescence, non-dispersive ultraviolet, calorimetric are good within the ranges shown. If air is used, O_2 will not exceed 21%, and should be well below 14% in the flue gas if you operate the system economically. CO_2 monitor ranges can be expanded to measure up to 21%, which might be necessary for burning something like a high quality cellulose. CO, ideally should be below 100 ppm. It would be ideal if nitrogen oxides are also below 100 ppm (reported as NO_2). SO_2 analyzers are shown with a wide range – 0–4,000 ppm. Burning coal that has roughly 2% sulfur will result in about 2000 ppm SO_2.

TABLE 12.4. Summary of Continuous Emission Monitors.

Pollutant	Monitor Type	Expected Concentration Range	Available Range*
O_2	Paramagnetic	5–14%	0–25%
	Electrocatalytic (e.g., zirconium oxide)		
CO_2	NDIR**	2–12%	0–21%
CO	NDIR	0–100 ppm	0–5000 ppm
NO_x	Chemiluminescent	0–4000 ppm	0–10,000 ppm
SO_2	Flame photometry	0–4000 ppm	0–5000 ppm
	Pulsed fluorescence		
	NDUV†		
SO_3	Colorimetric	0–100 ppm	
Organic compounds	Gas chromatography (FID)‡	0–50 ppm	
	Gas chromatography (ECD)§		
	Gas chromatography (PID)§§		
	IR absorption		
	UV absorption		
	GC/MS		

*For available instruments only. Higher ranges are possible through dilution.
**Non-dispersion infrared.
†Non-dispersion ultraviolet.
‡Flame ionization detector.
§Electron capture detector.
§§Photo-ionization detector.

12.6 SOME POSSIBLE SYSTEM COMPONENTS AND COSTS

The following are examples of two types of typical, currently available, extractive systems with costs in 1990 year dollars. Not all components listed are required for a system to be complete as several options are shown for some components (e.g., several sizes of probes are listed; only one is required).

12.6.1 EXTRACTIVE SYSTEM EXAMPLE A

Description	Price, 1990 $
Portable Probe System – 50 cm probe length – complete with 1/2% 30" per hour servo recorder, air purge and automatic stack exit correction	$15,200.00
Same as above – 100 cm probe length	$15,500.00
Opacity System – Includes transmitter, receiver, and control unit. 50-ft maximum span, severe environment rated	$3,950.00
Opacity System – Same as above, automatic zero and span system, 5-ft maximum span, EPA compliance	$6,700.00
Baghouse Monitor – Transmitter/receiver, retroreflector, and control unit, 5-ft span	$1,975.00
Multipoint baghouse system, double pass system – Adjustable set point with DPDT, 2A, relay contacts, external meter optional	$1,170/pt
Oxygen System, Zirconia, in situ – Control unit, O_2–1 V and 4–20 mA outputs. Meter scale .25–25% log output two decades, 2-ft stainless steel probe	$2,650.00
Oxygen/Combustible System, Extractive – Linear switchable scales, self-diagnostic and relay contacts	$5,950.00
Combustibles System, Extractive – Linear switchable scales	$3,300.00
Combustibles System, in situ – Control unit, 0–1 V and 4–20 mA outputs. Switchable scales. 2 ft S/S probe	$3,250.00
SO_2–NDIR–EPA – Sample conditioner, S/S probe, fiberglass case	$25,500.00
NO/NO_x, Chemiluminescent method – Auto Span System with sample conditioner, EPA compliance with NEMA 12 enclosure	$30,200.00

Opacity System, Single Pass – Dual beam for spans
up to 50 ft. Contains transmitter, receiver, control
box and lamp power supply, EPA compliance $12,950.00

Opacity System, Double Pass – Dual beam, for spans
to 185 ft. Contains transmitter, receiver, control box
and lamp power supply, EPA compliance $10,250.00

12.6.2 EXTRACTIVE SYSTEM EXAMPLE B

EPA gas analysis system to monitor oxygen, carbon monoxide, carbon dioxide and NO_x for an EPA gas analysis system made up of the following components:

(1) Stainless steel sample probe mounted on the stack
(2) Sample conditioning unit mounted on the stack
(3) Sample line connecting from the stack sample conditioning unit to the gas analyzer enclosures
(4) Gas analyzers, pumps, etc., housed in a stand-up enclosure for O_2, CO, CO_2, NO_x
(5) Solenoid valves that control the calibration gases may be in the stand-up enclosure or at the sample probe

Initially, the sample gas is extracted from the stack gas by a sample probe located on the stack. The sample is then "cleaned" by the sample conditioning unit, which is normally mounted on the smoke stack, as well. Thus, the sample first flows through the sample probe to a sample conditioning unit. Then, the sample is sent through the stainless steel sample line to the analyzer located in the customers control room. The calibration gas cylinders are used to calibrate the analyzers periodically to meet EPA requirements.

The following is a detailed description and price for an (EPA) system to monitor O_2, CO, CO_2 and NO_x.

Description	Price, 1990 $
Less than 2 ft, 1/2 in. stainless steel sample probe	$175.00

Sample conditioning unit containing the following:
a. Heated filter for particulate removal
b. Air operated chiller for water removal
c. Coalescing filter
d. Permapure dryer for additional water removal
to keep the dew point in the -20 to $-40°F$ region. System requires 4–6 CFM of plant air and a maximum 4 CFH of instrument air. Probe is back

purged at predetermined intervals. Housed in a fiber glass, NEMA 4 enclosure.	included in below analyzers
Electric heat traced line. 3/8 in. sample gas line. 1/4 in. cal. gas line. Insulated and jacketed weatherproof line, self-regulating includes signal cable between the sample conditioning unit and analyzer cabinet	$25.00/ft
Oxygen gas analysis system. This is an extractive system using a heated zirconia sensor. Contains coalescing inlet filter and vacuum pump.	$3,200.00
Carbon Monoxide gas analysis system meeting EPA requirements. The Carbon Monoxide CO) component in the sample gas is analyzed using the non-dispersive infrared (NDIR) technique. This method uses two cells, a sample and reference cell. A dual beam IR light source is projected through each cell and this energy is alternately shifted to a detector cell by a light chopper which operates at about 10 revolutions/second. The detector cell is the Luft type which acts as a mechanical variable capacitor. The cell is filled with the same CO concentration as the reference cell. The capacitance of the detector cell is a function of the CO concentration. The cell is part of LC (inductance, capacitance) resonant tank circuit which is connected to an oscillator. The frequency of oscillation is now a function of the concentration of CO gas in the sample stream. A frequency to voltage circuit converts the signal to a DC voltage.	$25,500.00
Carbon dioxide gas analysis system meeting EPA requirements. The CO_2 component of the gas stream is measured using a similar NDIR method as described for the CO system.	included in above CO price
The analyzer, pumps, automatic timing system for EPA calibration and all other controls and gauges are housed in a stand-up enclosure.	included in above analyzers
In the EPA system, the solenoid valves that control the calibration gases may be in the stand-up enclosure or at the sample probe.	included in above analyzers
NO/NO_x gas analysis system meeting EPA require-	

ments. This is an extractive analyzer using the chemiluminescent method. The stack gas sample is conditioned or "cleaned" of water and particulate matter by the sample conditioning unit before the sample is inserted into the analyzer. There are two methods of analysis. NO_x consists of Nitrogen Oxide and Nitrogen Dioxide. To measure NO_x, the sample gas flows through a temperature chamber (converter) that is heated to about 600°F. The Nitrogen Dioxide component is converted to NO, so that only NO flows into the combustion chamber. Ozone (O_3) is added to the NO in the combustion chamber, and a photomultiplier tube (PMT) measures the small amount of light emitted by the reaction of NO + O_3. To measure NO, the converter is bypassed and the sample gas stream goes straight into the combustion chamber, where it is mixed with the ozone. Again, the PMT measures the light emitted by the reaction. The Nitrogen Dioxide component is the difference between the NO_x and NO. The intensity of the emitted light, which is in the light wavelength region of 590–2705 nm, is proportional to the mass flow rate of NO in the combustion chamber. The voltage from the PMT is converted to a low impedance output voltage signal and a 4–20 mA interface. The PMT is controlled at 6°C by a thermoelectric cooler for stable operation. The ozone generator, which is internal to the instrument, uses an ultra-violet technique to generate ozone from an instrument air source. $28,050.00

This system does not include calibration gas cylinders or regulators. Not included

4 in., Felt Tip, Z-Fold, 3 Pen recorder $3,475.00

About the Authors

David L. Amrein is Vice President of Sales and Marketing. He oversees internal engineering, sales and marketing functions, and coordinates a worldwide system of manufacturers, representatives and licensees.

After college, Dave worked in design engineering related positions with Martin-Marietta. He then entered the air pollution control field with American Air Filter, where he spent seventeen years in project engineering electrostatic precipitator projects.

David has been with Fisher-Klosterman since 1984, in the business of custom dust collection collection and control equipment.

Richard Bundy has worked in air pollution control since 1967, primarily on fabric filter systems. For the last sixteen years, he has specialized in boiler and incinerator applications, where he has been responsible for supplying or upgrading the performance of over 100 installations. Dick's education was in civil engineering and business management at Miami University, the University of Idaho, and Indiana University.

His APC work started with Wheelabrator Corp. Then, in 1971, he started the fabric filter division of Standard Havens and was the technical and general manager until the division was sold to the United McGill Corp. in 1987, where he continued to be responsible for its operation.

In 1989, he started Bundy Environmental Technology to supply dry scrubber/fabric filter systems for industrial boiler and small incinerator applications. The company also provides consulting services for baghouse design and operational improvement.

Frank L. Cross, Jr., PE, DEE holds Bachelor of Science degrees in Chemical and Civil Engineering (Sanitary) and a Master of Engineering degree in Air Pollution Control from the University of Florida in Gainesville. He is a registered Professional Engineer in twelve states, including Florida, and is a Diplomate of Air Pollution Control in the American Academy of Environmental Engineers. Mr. Cross is also a certified Hazardous Materials Manager (CHMM).

Mr. Cross has over thirty years of diversified environmental engineering experience in industry and government. He was with the Air Program of the U.S. Environmental Protection Agency at Research Triangle Park, North Carolina; a Principal Consultant with Roy F. Weston, Inc.; Deputy Director of Operations for the Florida Department of Pollution Control; and President of Cross/Tessitore and Associates.

Presently, he is a principal consultant in Air with Harding Lawson Associates in Orlando, Florida.

Howard E. Hesketh, PE, Ph.D. DEE received B.S., M.S. and Ph.D. degrees in Chemical Engineering from The Pennsylvania State University. He was a Special Fellow at the Center for Air Environment Studies there.

Howard has industrial experience with DuPont, The Beryllium Corporation and Western Electric. He was a professor at The Pennsylvania State University and Kutztown State University and is currently Professor of Air and Hazardous Waste Management at Southern Illinois University. He is a consultant to industries, agencies and organizations.

Professional memberships include a former Director, Education Council Chairman and Vice President of the Air and Waste Management Association; a Diplomate and Examination Committee Chair of the American Academy of Environmental Engineers; and membership in ASME, AIChE, NSPE and ISPE.

Dr. Hesketh is the author of three textbooks and is co-author and editor of about twenty others. He has published numerous papers and was associate editor of *ASME Transactions Journals* and *Atmospheric Environment*.

John T. Quigley, PE, Ph.D. DEE earned B.S.Ch.E., M.S. and Ph.D. degrees in Environmental Engineering at the University of Wisconsin. His industrial background includes refinery experience with Shell Oil Company and chemical process engineering with Globe-Union, Incorporated. He conducted industrial effluent monitoring studies for Shell, so they could meet Los Angeles Country air and water pollutant emission standards.

Associated with University Extension since 1964, Dr. Quigley is currently Professor and Program Director for the Department of Engineering Professional Development. He is author and editor of an Environmental Engineering Series of Independent Study Courses in correspondence.

John is a registered Professional Engineer in the State of Wisconsin, with memberships in AWWA, WPCF, AIChE (Wisconsin Section), and Sigma Xi. He is a Diplomat of the American Academy of Environmental Engineers.

Frank W. Sherman, PE is Founder and President of Sherman Engineering, a consulting firm specializing in air pollution problems and issues. Mr. Sherman is a registered professional engineer in Illinois. He received a Bachelor of Science degree in Aerospace Engineering from Iowa State University, and a Master of Science degree in Mechanical Engineering from Southern Illinois University. He has over twenty-three years' experience in stationary and mobile source air pollution projects, with special emphasis on permitting and air quality planning issues. His clients include Fortune 500 companies and environmental agencies in several states, including Virginia, Texas, and Michigan. Prior to his current work, Frank was an engineer with the State of Illinois Environmental Protection Agency in Springfield, Illinois.

Patrick Walsh, JD received a Bachelor's degree in Electrical Engineering from Purdue University. He received Master's degrees in Nuclear Engineering and Environmental Engineering from the Massachusetts Institute of Technology (MIT). He was a consultant to the MIT-Harvard program in health studies and worked at Argonne National Laboratory before entering law school. In 1977, he received his Juris Doctor from the University of Wisconsin.

Mr. Walsh spent four years as an Assistant Attorney General in the Environmental Protection Unit of the Wisconsin Department of Justice. Then he joined the Madison law firm of Thomas, Parsons, Schaefer & Bauman. He currently is Co-Director of the Solid & Hazardous Waste Education Center and Associate Professor of Agricultural Engineering at the University of Wisconsin–Extension.

Pat has published a variety of articles on waste management. He has won a number of awards including Outstanding Teacher of the Year at the University of Wisconsin–Extension.

Lynda M. Wiese brings a wealth of practical experience to the business of air pollution control. As a field engineer for the Wisconsin Department of Natural Resources' (DNR) Southern District, she spent eight years piloting business and industry through the permitting and compliance process. An earth scientist by training, her strong background in air quality, permitting and plan review led her to a position of supervisor of the existing source permit unit for the DNR. She currently supervises the air management field operations for the Southern District of the DNR.

Lynda's knowledge of policy development issues dealing with permits and fees has been invaluable to the Wisconsin Clean Air Act task force — an advisory body of industry and environmental groups reviewing state laws for compliance with the Act. Lynda also serves as chair of the DNR's Asphalt Paving Technology Team. This is a cross-media team that interacts with the paving industry on a variety of environmental concerns.

Index